13th International Gas Analysis Symposium (GAS 2026)

EPJ Web of Conferences Volume 352 (2026)

Versailles, France
27-29 January 2026

Editors:

Jérôme Lopez
Lauryanne Teulon

ISBN: 979-8-3313-3277-8

Printed from e-media with permission by:

Curran Associates, Inc.
57 Morehouse Lane
Red Hook, NY 12571

Some format issues inherent in the e-media version may also appear in this print version.

For additional information, please contact EDP Sciences – Web of Conferences
at the address below.

EDP Sciences – Web of Conferences
17, Avenue du Hoggar
Parc d'Activité de Courtabœuf
BP 112
F-91944 Les Ulis Cedex A
France

Phone: +33 (0) 1 69 18 75 75
Fax: +33 (0) 1 69 28 84 91

contact-edps@webofconferences.org

Additional copies of this publication are available from:

Curran Associates, Inc.
57 Morehouse Lane
Red Hook, NY 12571 USA
Phone: 845-758-0400
Fax: 845-758-2633
Email: curran@proceedings.com
Web: www.proceedings.com

TABLE OF CONTENTS

13th International Gas Analysis Symposium

January 27 to 29, 2026,

Paris - Porte de Versailles, France

In partnership with HYVOLUTION

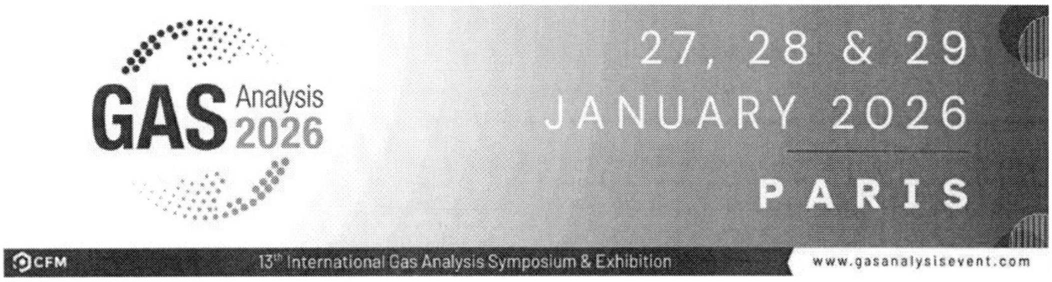

GAS Analysis is a biennial event dedicated to gas analysis, launched in 1999. It was organized from its origin in the Netherlands by the Dutch Standards Institute (NEN) and ISO/TC 158 "Analysis of gases" until 2021.

Since 2022, the congress has been hosted in France and is organized by the Collège Français de Métrologie (CFM).

The 2022 edition was held within the Global Industrie trade show, as part of Measurement World. Since 2024, it has been hosted by the Hyvolution trade show, dedicated to hydrogen. The organisation is supported by an Organising Committee and a Scientific Committee consisting of experienced persons covering the various fields of gas analysis. The Organising Committee is involved in the overall organisation of GAS Analysis event whereas the Scientific Committee is responsible for the symposium programme and poster session by evaluating the abstracts submitted and compiling the symposium programme.

This 13th edition offers another dynamic 3-day program, featuring parallel sessions tackling global challenges. 100% in English, with over 100 presentations covering oral and poster sessions, along with a round tables, a short course and a variety of social events, GAS Analysis symposium continues to be a hub of cutting-edge discussions and networking opportunities.

FULL PROGRAM ACCESS:

https://www.gasanalysisevent.com/images/GAS26/Programme/GAS2026_Full_Programme.pdf

ORGANIZING COMMITTEE MEMBERS

Chair: Etienne Smith - MIRO Analytical / Switzerland

João ALVES E SOUSA - IPQ & EURAMET / Germany

Annarita BALDAN – VSL / The Netherlands

Paul BREWER - NPL / United Kingdom

Ortwin COSTENOBLE - PAC PETROLEUM ANALYZER COMPANY / The Netherlands

Jack DE JONG - AIR LIQUIDE / The Netherlands

Maxime DUFOND - ENGIE / France

Philippe FAYOLLE - ENVEA / France

SCIENTIFIC AND TECHNICAL COMMITTEE MEMBERS

Chair: Annarita BALDAN – VSL / The Netherlands

Etienne BASSET - ENGIE / France

Paul BREWER - NPL / United Kingdom

Kevin CLEAVER - BOC (THE LINDE GROUP) / United Kingdom

Ortwin COSTENOBLE - PAC PETROLEUM ANALYZER COMPANY / The Netherlands

Jack DE JONG - AIR LIQUIDE / The Netherlands

Tracey JACKSIER - AIR LIQUIDE / USA

Heinrich KIPPHARDT - BAM BUNDESANSTALT FÜR MATERIALFORSCHUNG UND - PRÜFUNG / Germany

Wouter KNOL - SHELL Technology / The Netherlands

Tatiana MACÉ - LNE / France

Stefan PERSIJN - VSL / The Netherlands

Joey WALKER - EFFECTECH / United Kingdom

PARTNERS

1) SPONSORS

2) HOSTED BY

3) ORGANIZERS

4) MEDIA & OTHER PARTNERS

Statement of Peer review

In submitting conference proceedings to *Web of Conferences*, the editors of the proceedings certify to the Publisher that

1. They adhere to its **Policy on Publishing Integrity** in order to safeguard good scientific practice in publishing.
2. All articles have been subjected to peer review administered by the proceedings editors.
3. Reviews have been conducted by expert referees, who have been requested to provide unbiased and constructive comments aimed, whenever possible, at improving the work.
4. Proceedings editors have taken all reasonable steps to ensure the quality of the materials they publish and their decision to accept or reject a paper for publication has been based only on the merits of the work and the relevance to the journal.

Title, date and place of the conference

13th International Gas Analysis Symposium
January 27 to 29, 2026,
Paris - Porte de Versailles, France
In partnership with HYVOLUTION

Proceedings editor(s):

Jérôme LOPEZ, Collège Français de Métrologie
Lauryanne TEULON, Collège Français de Métrologie

Date and editor's signature

February 2nd 2026

Online monitoring of energy content and impurities in hydrogen using a micro gas chromatograph

Eva Krolis [1*], and *Füsun* Karaburun-Yavuz[1]

[1]Qmicro B.V., Neptunusstraat 21, 7521 WC, Enschede, The Netherlands

Abstract. Hydrogen is increasingly considered as a clean alternative to natural gas, creating a need for reliable online measurements of its energy content and impurity concentrations. This study evaluates the performance of the DynamiQ-X micro gas chromatograph (GC) for monitoring factory-grade hydrogen. Helium and hydrogen were used as carrier gases in a two-unit configuration, measuring in parallel. Gas mixtures were prepared according to frequently cited impurity limits, and the system's repeatability, accuracy and detection limits were evaluated. The measured calorific value and Wobbe index met market requirements, while the component repeatabilities were high and primarily determined by the concentration. The limits of detection (LOD) and quantification (LOQ) were low for all components, with minimal differences between both carrier gases. Notably, an oxygen LOQ below 1 ppm was achieved when hydrogen was used as the carrier gas. The results demonstrate that the equipment is suitable for online monitoring of factory-grade hydrogen and can achieve consistent and reliable analyses.

1 Introduction

As the energy sector pushes toward net zero, hydrogen is gaining momentum as a clean alternative to natural gas. While blending up to 20% hydrogen into natural gas networks serves as a transitional step, the long-term goal is a dedicated hydrogen infrastructure, serving industrial, commercial and even domestic consumers [1]. This shift introduces new challenges, one of which is in accurately measuring the energy content and impurities. Industry-grade hydrogen with minimum purification degrees of 98% or 99.5% can be used in factories and distribution through a pipeline system [1-3]. To ensure safe and efficient use, online monitoring of the calorific value and Wobbe index is essential, alongside precise measurement of the hydrogen content and impurities [4].

The DynamiQ micro Gas Chromatograph (GC) is equipped with several MEMS micro-machined chip components, which are connected by a patented chip to chip connection technology. This architecture enables a compact instrument footprint in an explosion proof design. The GC operates in a backflush-to-detector configuration, which increases the lifetime of the columns and reduces the total analysis time to typically less than a minute [5].

This study evaluates the performance of the DynamiQ micro GC using impurity thresholds from standards and market requirements. Helium and hydrogen carrier gas are compared, where helium allows for hydrogen quantification and hydrogen carrier gas improves the oxygen detection limit. Accuracy and repeatability of both carrier gases are compared and assessed across a range of gas mixtures and evaluated against the international recommendation OIML R140 [6]. Additionally, limits of detection (LOD) and quantification (LOQ) are determined following the methods specified in ISO 21087:2019 [7].

2 Materials and methods

Standards and market requirements for industry-grade hydrogen are still under development. The impurity types and levels as well as the minimum purification degree of either 98% or 99.5% remain under active discussion. ISO 14687:2025 [4] specifies the minimum quality characteristics of hydrogen fuel for residential, commercial, industrial, vehicular and stationary applications. While the focus has mostly been on Grade D, regarding fuel cell applications, grade A specifies industrial-grade limits intended for internal combustion engines. The technical specification CEN/TS 17977:2023 [1] defines hydrogen quality to be distributed by rededicated natural gas infrastructure, aiming to balance the requirements for the producer, grid operators and end-users. Additionally, the European Association for Streamlining of Energy Exchange has developed a Common Business Practice [2] for hydrogen in the natural gas grid, leading to a proposed specification. In 2023, DNV and KIWA conducted a study [3] for the Dutch Ministry of Economic Affairs and Climate Policy, recommending a minimum hydrogen purity of 99.5%, in contrast to other studies.

While these standards and recommendations largely agree on the limits of impurities, some differences remain. The most frequently cited values are

* Corresponding author: eva.krolis@qmicro.com

 https://doi.org/10.1051/epjconf/202635201001

summarized in Table 1. For oxygen, the limit is 0.1 mol% in grids with no exit points to underground gas storages or sensitive customers, otherwise a maximum of 10 ppm is allowed. All parameters up to and including carbon dioxide can be analyzed using the DynamiQ micro GC.

Table 1. Summary of standards and recommendations for industry-grade hydrogen.

Parameter	**Limit**
Hydrogen	≥98% (or ≥99.5%)
Wobbe index	42 – 46 MJ/m^3
Inerts (N2, Ar, He)	≤2%
Hydrocarbons	≤2%
Oxygen	≤0.1 % or ≤10 ppm
Carbon monoxide	≤20 ppm
Carbon dioxide	≤20 ppm
Water	≤250 or ≤60 ppm
Water dewpoint	≤-8 °C at 70 bar
Hydrocarbon dewpoint	≤-2 °C at 1-70 bar
Total sulphurs	≤7 ppm
Ammonia	≤13 ppm
Halogenated compounds	≤0.05 ppm

Both helium and hydrogen were used as carrier gases for the DynamiQ micro GC. Helium grade 5.0 was used, while hydrogen was generated on-site using a hydrogen generator, providing comparable purity. The generated hydrogen also served as the balance gas for preparing mixtures with the desired impurities, as depicted in Figure 1. For each impurity, at least five evenly spaced concentrations were prepared to assess linearity. Additionally, measurements near 1 ppm were performed to determine the limit of detection (LOD) and limit of quantification (LOQ). Finally, a composition representative of industry-grade hydrogen was measured to evaluate the calculation of calorific value and Wobbe index.

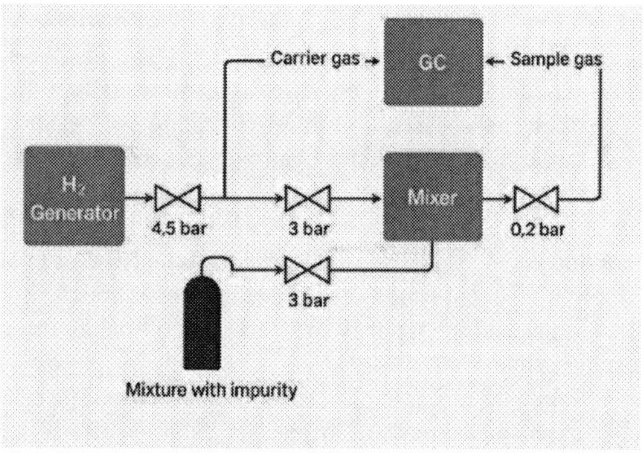

Fig. 1. Measurement setup.

The DynamiQ-X can contain up to four GC units that each perform in parallel a different GC analysis under individually optimized method conditions. They consist of an injector, columns and thermal conductivity detectors (TCDs), as depicted in Figure 2. For these analyses, only two GC units were needed where the same setup could be used for both helium and hydrogen carrier gas. The instrument consists of a stream selector with 4 sample streams. One sample stream can be measured at the same time but the two GC units analyze this gas simultaneously with parameters optimized for different components.

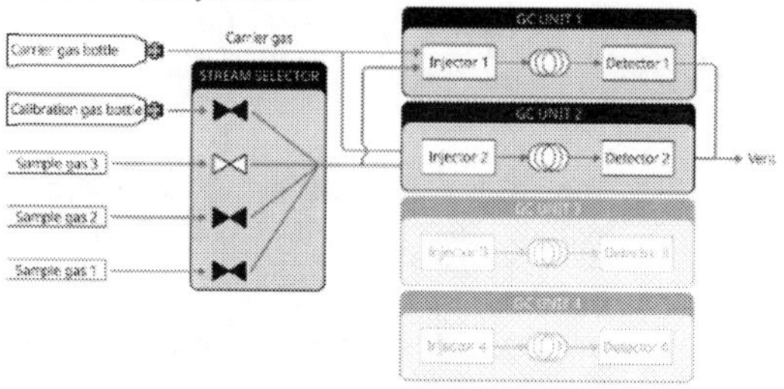

Fig. 2. Schematic of the stream selector and GC units.

The DynamiQ-X is designed for continuous monitoring and works therefore stand-alone. The generated data can be communicated via Modbus and the instrument is WELMEC 7.2 compliant [8, 9]. The ATEX/IECEx certified analyzer is housed in an explosion-safe marine-grade aluminium IP65 enclosure. The instrument has a volume of 10 L and a low carrier gas consumption of one gas bottle for over one year [5].

The two GC units consist of an injector and two detectors integrated on a chip, a pre-column and an analytical column. This configuration enables the use of backflush technology, which decreases the analysis time and prevents late-eluting or unwanted compounds from entering the analytical column. During sample injection, the system operates in foreflush mode. At a defined point in the analysis, the flow through the pre-column is reversed and directed to the backflush detector (TCD BF), where a combined peak of all late-eluting components is recorded. Simultaneously, components that have already reached the analytical column continue to elute at the foreflush detector (TCD FF). A schematic is given in Figure 3.

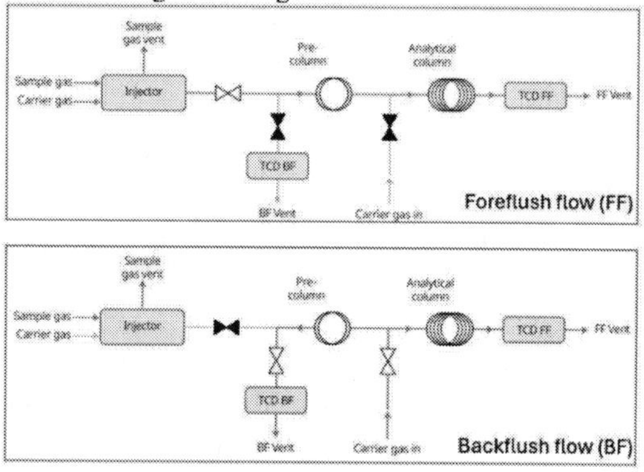

Fig. 3. Schematic of foreflush (FF) and backflush (BF) flow.

The thermal conductivity detectors (TCDs) operate by detecting changes in heat transfer from a heated filament as different components pass through. The magnitude of the signal is determined by the difference in thermal conductivity between the carrier gas and the analyte. Helium and hydrogen are commonly used carrier gases because of their high thermal conductivities. Hydrogen has a lower viscosity than helium, resulting in shorter retention times. Consequently, analyses with hydrogen carrier gas are faster, with an analysis time of 35 seconds compared with 45 seconds with helium carrier gas.

3 Results and discussion

3.1 Chromatograms

The chromatograms using hydrogen as the carrier gas are shown in Figure 4, with a total analysis time of 35 seconds. The first GC unit is configured to measure oxygen, nitrogen, methane and carbon monoxide. As indicated in Table 1, the concentrations of oxygen and carbon monoxide are expected to be significantly lower than those of nitrogen and methane. Nevertheless, all components exhibit sufficiently well-defined peak shapes for reliable integration.

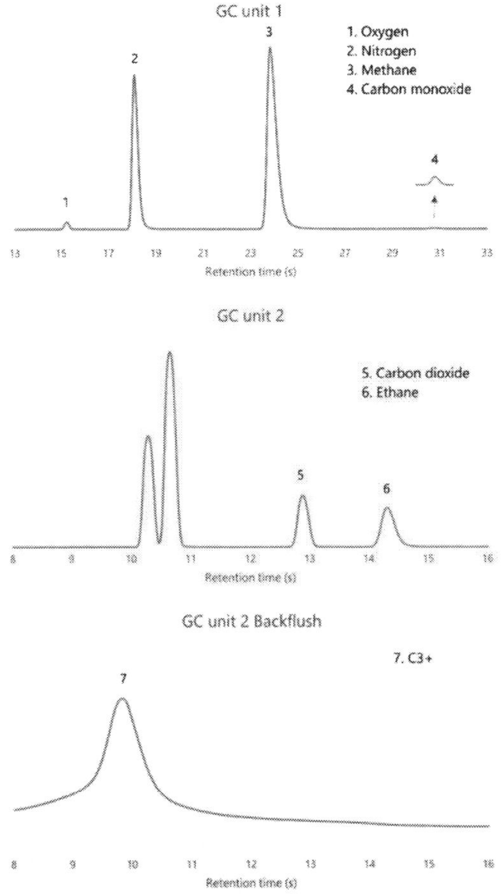

Fig. 4. Example chromatograms using hydrogen carrier gas

Since the sample gas is directed to both GC units simultaneously, the same component types appear in the chromatogram of GC unit 2. However, they are neither separated nor quantified there. On the foreflush detector of GC unit 2, carbon dioxide and ethane are measured, while hydrocarbons from propane and higher (C3+) are detected on the backflush detector. During post-processing, ethane and the C3+ hydrocarbons are combined and reported as C2+, as higher hydrocarbons are only expected to occur in low amounts. Due to limitations in available sample gases, a mixture containing high amounts of higher hydrocarbons up to n-hexane had to be used. Despite this, the peak shape is still good, only slightly broader, indicating that the method works well even with heavier hydrocarbons that are not expected in real-life samples.

The chromatograms using helium as the carrier gas are shown in Figure 5, with a total analysis time of 45 seconds. The same components are measured on both GC units as when hydrogen is used as the carrier gas. In addition, the hydrogen concentration can be quantitatively determined on GC unit 1 when helium carrier gas is used. Because hydrogen has a higher thermal conductivity than helium, this results in a negative peak in the chromatogram. Despite the large hydrogen peak eluting before oxygen, the separation between hydrogen and oxygen is good.

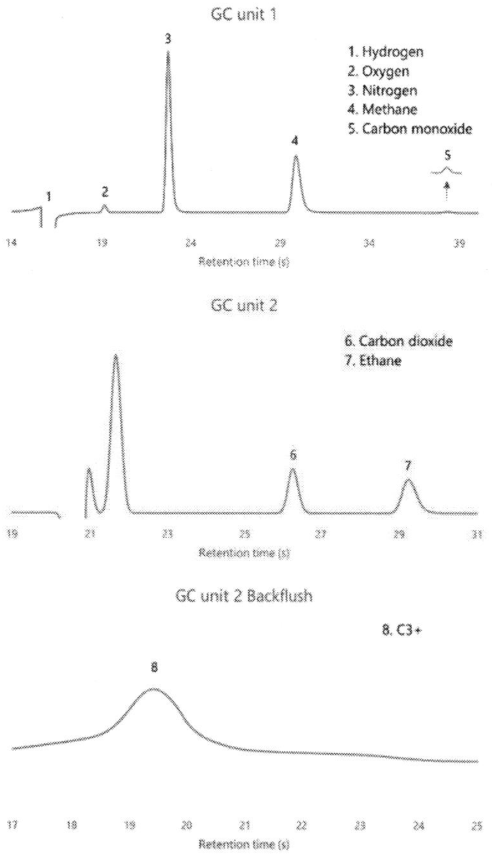

Fig. 5. Example chromatograms using helium carrier gas

3.2 Repeatability

The repeatability was evaluated as the relative standard deviation (RSD) of the concentrations between 10 consecutive measurements, where the results are presented in Table 2. The RSDs were determined at the maximum allowed concentration limits from the assessed standards and recommendations. For impurities present at percentage levels, the RSDs are well below 0.5%. For carbon dioxide, carbon monoxide and oxygen, with a limit of 10 or 20 ppm, the RSDs are higher due to the lower signal. The repeatability of carbon dioxide and of oxygen measured with hydrogen carrier gas remain below 1%. When oxygen is measured with helium carrier gas, the repeatability is slightly higher (below 2%), as the large hydrogen peak affects the local baseline. Carbon monoxide shows the highest spread, with RSDs below 3% for both carrier gases. This is attributed to its longer retention and low concentration, which leads to a broader, flatter peak that is more challenging to integrate consistently.

Table 2. Relative standard deviation (RSD) of the concentrations

Component	Concen-tration	Relative standard deviation (RSD)	
		Helium carrier	Hydrogen carrier
Methane	2%	0.10%	0.05%
C2+	2%	0.08%	0.15%
Nitrogen	2%	0.03%	0.03%
Carbon dioxide	20 ppm	0.63%	0.58%
Carbon monoxide	20 ppm	2.30%	2.89%
Oxygen high range	0.1%	0.14%	0.08%
Oxygen low range	10 ppm	1.62%	0.57%

3.3 Calorific value and Wobbe index

The calorific value and Wobbe index were determined in accordance with ISO 6976:2016 [10], as described in literature [1, 3]. A combustion temperature of 15 °C, a metering temperature of 15 °C and a metering pressure of 101.325 kPa were used. The relative error of the average measurement and the repeatability, expressed as the RSD of 10 consecutive measurements, are given in Table 3 and Table 4. The relative error of the calorific value is well below 0.50%, which complies with the requirement of accuracy class A in international recommendation OIML R140 [6]. Although OIML R140 is intended for fuel gases such as natural gas, its performance criteria can be applied to factory-grade hydrogen as well in absence of a dedicated metrological recommendation.

Table 3. Average measurement, relative error and relative standard deviation (RSD) of 10 consecutive measurements of the calorific value

	Helium carrier	Hydrogen carrier
Average (MJ/m^3)	12.430	12.442
Relative error	0.064%	0.023%
RSD	-0.009%	0.091%

Table 4. Average measurement, relative error and relative standard deviation (RSD) of 10 consecutive measurements of the Wobbe index

	Helium carrier	Hydrogen carrier
Average (MJ/m^3)	42.6413	42.629
Relative error	0.028%	0.009%
RSD	0.010%	-0.020%

3.4 Limits of detection (LOD) and quantification (LOQ)

To determine the limits of detection (LOD) and quantification (LOQ), measurements were performed at impurity concentrations around 1 ppm, as this is close to the expected detection limit. The first method to determine the LOD is to identify the lowest analyte concentration that can be reliably distinguished from the background noise. In gas chromatography, a signal-to-noise ratio of 3 is generally accepted as the criterion for the instrument limit of detection:

$$LOD = \frac{3 \times Noise}{RF} \quad (1)$$

where *Noise* is the RMS noise and *RF* is the response factor of the impurity. The response factor is obtained from the calibration curve, by dividing the measured peak area by the analyte concentration. The instrument limits of detection are summarized in Table 5.

Table 5. Instrument limit of detection, based on the noise.

Component	LOD based on noise (ppm)	
	Helium carrier	Hydrogen carrier
Methane	0.08	0.08
C2+	0.07	0.35
Carbon dioxide	0.04	0.08
Carbon monoxide	0.19	0.21
Oxygen	0.06	0.02

Following the definitions in ISO 21087 [7], this represents the instrument detection limit. The method detection limit, which takes the whole measurement procedure into account, is calculated as:

$$x_{LOD} = 3 \times s_0' \quad (2)$$

where s_0' is the standard deviation of 10 consecutive measurements. The method limits of detection are shown in Table 6. For all components except carbon monoxide, the LOD is below 1 ppm. Carbon monoxide elutes late on the first GC unit, resulting in a flatter peak shape that is more difficult to integrate and slightly increases the standard deviation.

Table 6. Method limit of detection, based on the standard deviation.

Component	LOD based on s_0' (ppm)	
	Helium carrier	Hydrogen carrier
Methane	0.36	0.43
C2+	0.52	0.78
Carbon dioxide	0.18	0.20
Carbon monoxide	1.04	1.21
Oxygen	0.33	0.08

The limit of quantification is the lowest concentration of an impurity that can be quantified with acceptable precision and accuracy. It is calculated as:

$$x_{LOQ} = k_Q \times s_0' \quad (3)$$

where k_Q depends on the maximum allowed concentration in the specification. In this study, $k_Q = 10$ for all impurities.

The limits of quantification are given in Table 7. All quantification limits are below 5 ppm. Generally, components that elute earlier exhibit lower LOQs, primarily due to a better peak shape. Differences between helium and hydrogen as carrier gas are small for most components. The only notable exception is oxygen, where hydrogen carrier gas achieves an LOQ below 1 ppm. This improved performance is caused by good separation of oxygen from other components and its short retention time, resulting in a sharp peak.

Table 7. Limit of quantification, based on the standard deviation.

Component	LOQ based on s_0' (ppm)	
	Helium carrier	Hydrogen carrier
Methane	1.21	1.43
C2+	1.72	2.61
Carbon dioxide	0.58	0.65
Carbon monoxide	3.48	4.04
Oxygen	1.10	0.25

4 Conclusions

This study evaluated the performance of the DynamiQ-X for measurements of energy content and impurities in factory-grade hydrogen. The system operated reliably with both helium and hydrogen as carrier gases, without requiring hardware modifications. The most frequently cited limits for impurities were summarized from literature and used to prepare sample gases. The measured calorific value and Wobbe index met market requirements, and the repeatabilities of the impurity measurements were high, governed primarily by the impurity concentration. The limits of detection and quantification were low, with only minor differences between the two carrier gases. Notably, an oxygen LOQ below 1 ppm was achieved when hydrogen was used as the carrier gas.

Overall, the results indicate that the equipment is suitable for online monitoring of energy content and impurities in factory-grade hydrogen. As hydrogen quality monitoring becomes increasingly important for both regulatory compliance and process control, the DynamiQ-X provides a robust basis for achieving consistent and traceable measurements.

References

1. European Committee for Standardization (CEN), CEN/TS 17977:2023 Gas infrastructure – Quality of gas – Hydrogen used in rededicated gas systems (2023)
2. EASEE-gas Quality Harmonisation Working Group, Common Business Practice on Hydrogen (2022)
3. E.A. Polman, M.L. Doelman, S. van Greuningen, J.W. Turkstra, M. van Essen, H. Vlap, A follow-up study into the hydrogen quality requirements, KIWA Technologies and DNV (2023)
4. International Organization for Standardization (ISO), ISO 14687:2025 Hydrogen fuel quality – Product specification (2025)
5. Qmicro by Sensirion, DynamiQ-X process GC – Gas chromatographs for on-line gas analysis (2024)
6. International Organization of Legal Metrology, International Recommendation OIML R140 – Measuring systems for gaseous fuel (2007)
7. International Organization for Standardization (ISO), ISO 21087:2019 Gas analysis — Analytical methods for hydrogen fuel — Proton exchange membrane (PEM) fuel cell applications for road vehicles (2019)
8. European Cooperation in Legal Metrology (WELMEC) e.V., WELMEC Guide 7.2 – Software Guide (EU Measuring Instruments Directive 2014/32/EU) – Version 2022 (2022)
9. Qmicro by Sensirion, Application note – Natural gas and natural gas blended hydrogen (2025)
10. International Organization for Standardization (ISO), ISO 6976:2016 Natural gas – Calculation of calorific values, density, relative density and Wobbe indices from composition (2016)

Integrated Analytical Measurement Systems for Safe and Efficient Green Hydrogen Production

Thorsten Reisinger [1*]

[1]AGT-PSG GmbH, 65239 Hochheim am Rhein, Germany

Abstract. Hydrogen, which is produced from renewable energies, is of fundamental importance in making industry climate-neutral. Despite technical advances and political support, there are still a number of challenges regarding the safety, reliability, efficiency and cost-effectiveness of green hydrogen production facilities. This article deals with the use of process analyser systems in hydrogen production and purification. In particular, the focus is on the implementation of process analytics in the plant, its significance for plant safety and hydrogen quality, and aspects of sample conditioning to ensure the functionality of process analysers in electrolysis plants, specifically using alkaline and proton exchange membrane (PEM) technologies. It emphasises the importance of powerful, fully-integrated process analyser systems. This guarantees stable processes, minimises downtime and ensures product quality. In turn, it contributes to an increasing market ramp-up of green hydrogen technologies.

1 Introduction

In order to replace fossil fuels with low-carbon alternatives, it is becoming increasingly necessary to drive forward the transition to greenhouse gas-neutral energy systems. In this context, hydrogen plays an important role due to its high energy density, its potential as a raw material, fuel and energy storage medium, and its versatility. Hydrogen obtained by electrolysis of water from renewable sources is usually considered ‘green hydrogen’. Its production represents an opportunity to significantly reduce greenhouse gas emissions in various industries, such as the chemical industry, the refinery and steel sectors, as well as transport. However, a hydrogen-based energy system still faces technical and economic challenges, despite its significant market potential.

Recent delays and cancellations of projects have highlighted how sensitive green hydrogen initiatives are to predictable capital expenditure, operating costs and high system reliability [1-5]. The safety risks associated with the production of hydrogen and oxygen also necessitate strict control and monitoring strategies. In the past, inadequate measures to ensure gas purity and monitor cross-contamination have led to serious industrial accidents.

The rupture of an oxygen separator drum in a potassium hydroxide electrolysis plant in the United Kingdom in 1975 demonstrated the catastrophic consequences that inadequate analytical monitoring can have. Such events underscore the need for continuous, accurate and reliable process analytics as a fundamental component of hydrogen production facilities. This article examines the technical requirements and integration strategies for analytical measurement systems in green hydrogen production and emphasises their role in ensuring safety, operational stability and product quality.

2 The Role of Process Analytics in Water Electrolysis

2.1 Electrolysis Technologies and Operating Conditions

At the present time, alkaline electrolysis (AEL) and proton exchange membrane electrolysis (PEM) are the most mature and widely used technologies for the production of renewable hydrogen. Despite the fact that both systems are predicated upon the electrochemical splitting of water into hydrogen and oxygen, there are significant differences between them in terms of electrolyte composition, membrane design, operating pressure and dynamic behaviour.

PEM electrolysis is distinguished by high current densities, a compact design and a rapid response to fluctuating electrical input power. It is therefore especially good for renewable energy sources. AEL systems are stable over a long period of time and are cheaper, but slower and have lower current densities. Whatever technology is used, both systems produce hydrogen and oxygen very close to each other. These two gases are kept apart only by membranes or diaphragms. Consequently, issues such as membrane wear, pressure imbalances or operational malfunctions have the potential to result in the cross-diffusion of gases between the anode and cathode chambers.

* Corresponding author: tre@perfectsamplegas.de

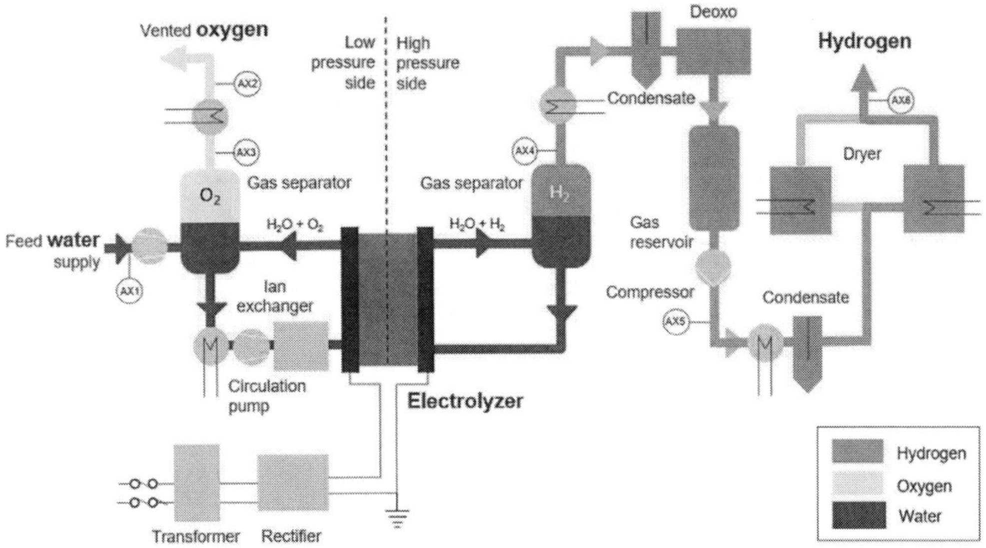

Fig 1. Simplified system design using the example of the PEM electrolysis system including analyser measuring points (AX1-AX5) [6]

2.2 Safety-Relevant Analytical Measurements

Due to safety considerations, it is mandatory to undertake continuous measurement of the oxygen content in hydrogen (O_2 in H_2) and the hydrogen content in oxygen (H_2 in O_2). It is a well-established fact that these parameters often serve as indicators of membrane integrity; moreover, they are frequently incorporated as shutdown or kick-out criteria in safety-oriented systems (SIS). This point is further illustrated by the tags AX-3 and AX-4 in Table 1.

Although the usual crossover concentrations are much lower than the lower explosion limits, their detection enables early warning of abnormal operating conditions. But unstable or overly sensitive measurement systems can lead to false alarms and unnecessary plant downtime. This means that the plants are less available and perform less well economically.

Analytical systems must therefore be highly accurate and respond quickly while ensuring that signals remain stable over time in demanding conditions.

3 Hydrogen Purification and Quality Control

3.1 Hydrogen Quality Requirements

The level of purity required for hydrogen is dependent on its intended use. Hydrogen that is to be used in fuel cells or for chemical synthesis must meet strict specifications, whereas certain industrial applications tolerate moderate levels of impurities. International standards such as ISO 14687 specify maximum permissible concentrations for a variety of impurities.

Typical impurities in hydrogen include nitrogen, oxygen, water vapour, carbon monoxide (CO) and, in some cases, carbon dioxide, sulphur compounds, ammonia and hydrocarbons. Even in trace amounts, some components, particularly CO and sulphur compounds, have the potential to cause irreversible

Tag	Measuring Point (Sampling Point)	Measurement Objective	Measuring Components	Common Process Analyzer Principle
AX-1	Feed water supply	Quality	Conductivity pH	Sensor, electrical resistance Electrochemical sensor, H^+ Ions
AX-2	O_2 pathway after gas separation	Quality	H_2, N_2, CO_2 in O_2	Gas chromatography
AX-3	Electrolyser, outlet after gas separation - O_2 pathway	Safety	O_2 in H_2	Continuous gas analyser (CGA), paramagnetic
AX-4	Electrolyser, outlet after gas separation - H_2 pathway	Safety	H_2 in O_2	CGA, thermal conductivity
AX-5	H_2 pathway after compressor	Process optimization	O_2, in H_2	CGA, paramagnetic
AX-6	H_2 pathway after gas separation	Quality	O_2 in H_2 Moisture in H_2 O_2, N_2, CO in H_2	CGA, electrochemical Sensor, capacitive silicon Gas chromatography

Tab. 1. Measuring points and common process analyser technology

damage to catalysts or stop fuel cells from working properly.
It is therefore essential to supplement hydrogen purification systems with sophisticated analytical measurement systems that are capable of detecting impurities at levels as low as a few parts per million (ppm ranges) or even parts per billion (ppb ranges).

3.2 Analytical Techniques for Purity Monitoring

In order to monitor the purity of hydrogen and oxygen, it is common to combine various analytical methods, such as continuous gas analysis, process gas chromatography and trace moisture analysis.
Paramagnetic oxygen analysers are widely used in industry due to their high levels of selectivity, stability and suitability for safety-critical applications. Thermal conductivity detectors (TCDs) offer a robust method for measuring hydrogen, but are sensitive to moisture and the composition of the background gas.
In the hydrogen purification and processing stages, process gas chromatographs are vital for multi-component analysis and product specification verification. Additional measurements of water quality, such as pH and conductivity, are also critical to protect membranes and electrodes from damage.

4. Sampling and Sample Conditioning Challenges

4.1 Process Environment Constraints

In hydrogen production plants [1], sampling systems must be able to operate under extreme conditions, such as high temperatures (< 90°C), high pressure (< 80 barg) and water-saturated gas streams. Furthermore, there are many sampling points located in areas that are at risk of explosion (Ex zones). This means that further restrictions are imposed on the design of the system and the selection of materials. If these requirements are not properly taken into account, this can lead to condensation, pressure fluctuations and contamination. Ultimately, this affects the accuracy of the measurement and the response time.

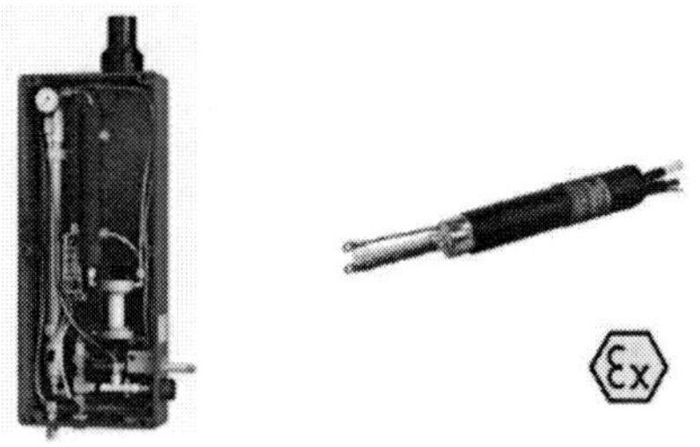

Fig. 2. Sample extraction probe (left) and heated sample line (right) – Robust product design (PSG) to secure the reliable sample transport to the analyser

4.2 Primary and Secondary Sample Conditioning

The objective of primary sample preparation is to ensure that the process gases are suitable for transport and analysis. To this end, pressure must be reduced, the flow stabilised and condensation prevented (Fig. 2). It is important to note that if pressure is not reduced correctly, this can lead to cooling effects and water precipitation. This has a particularly negative effect on thermal conductivity measurements. Secondary sample preparation, which is normally located close to the analyser, allows precise regulation of pressure, temperature and flow. The system provides functions such as particle filtration, moisture removal, sample switching and calibration. Precise control of the dew point is essential for TCD-based measurements, as water vapour significantly influences the thermal conductivity signals. The setup shown in Fig. 3 enables reliable and flexible secondary sample preparation for the production of green hydrogen.

Fig. 3. Secondary sampling system in green H_2 production

The process has multiple sampling points, which allow representative samples to be taken from different locations. All sampling lines are either frost-protected or heated, which helps to ensure stable conditions. Sample switching blocks enable efficient forwarding to multiple analysers. Integrated sample preparation includes calibration and validation, which helps to ensure accurate and consistent measurements. The system is compatible with various analysers, including paramagnetic sensors for O_2 in H_2 and TCD and paramagnetic sensors for H_2 in O_2. Utilising robust, proven TCD technology, this concept facilitates straightforward validation and calibration, minimises wear and tear, and thus ensures high reliability and long-term operation.

5. Analyzer Selection and Safety Integrity Considerations

When selecting analysers, it is necessary to take into account their measurement performance, robustness, maintenance requirements and compatibility with safety integrity level (SIL) requirements. A common practice is to combine paramagnetic and TCD analysers in order to utilise the complementary strengths of both devices. Dynamic cross-correction of TCD signals using paramagnetic oxygen measurements enables accurate hydrogen detection in advanced system architectures, even when the background gas conditions are extreme. This enables hydrogen measurement in the oxygen (O_2) path at the electrolyser in both normal operating conditions (i.e. H_2 in O_2) and during periods of start-up or malfunction (i.e. H_2 in N_2 or H_2 in an O_2/N_2 mixture). Configurations of this type have been shown to increase measurement reliability and functional safety. They guarantee fail-safe operation without excessive conservatism. Integrated gas coolers guarantee the required stable dew point of the sample.

Operating Condition	Typical H_2 Level in O_2	Interpretation / Notes
Normal crossover operation	< 0.1 % (typically a few 1.000 ppm)	Normal membrane permeation; safe operating range.
Pre-alarm (≈ 25% of LEL)	~ 1.0 - 1.5 % vol H_2	Early warning threshold; indicates increasing crossover or membrane degradation.
Main alarm/ Trip (≈ 50% of LEL)	~ 2.0 - 2.5 % vol H_2	Automatic shutdown to prevent approach-ing flammable conditions. Strong indication of membrane wear or failure.
Lower Explosion Limit (LEL)	~ 4.0 % vol H_2 in O_2	Mixture becomes flammable; must never be reached under operating conditions.

Fig. 4. Hydrogen crossover safety threshold in electrolysers (Typical industry value).

6. System Integration and Infra-structure

Analysis systems are generally installed in special cabinets or containers, particularly in the case of hydrogen purification plant sections which are located outside the electrolysis cell rooms. When constructing these enclosures, it is essential to take into account environmental aspects, the classification of hazard zones, ventilation and accessibility for maintenance work.

Fig. 5. Analysis cabinet with mesh doors (open design) for safety-focused online analysis at the electrolyser

Depending on the safety concept and system integration requirements, a few options like open (mesh doors) or closed cabinets as well as containers, especially for the integration of process gas chromatographs, are available.

Furthermore, it is crucial to have a dependable supply infrastructure, including carrier gases, instrument air, and electricity, to maintain continuous analysis operations. To guarantee the long-term reliability of the system, digital interfaces and predictive maintenance features are progressively being integrated into analysis systems.

7. Discussion and Outlook

The successful use of green hydrogen technologies depends on how efficient electrolysers are and how much electricity from renewable sources is available. It also depends on how reliable and safe auxiliary systems are. Analytical measurement systems are very important here because they can spot faults early, prevent dangerous situations and make sure products meet quality requirements.

It is important to note that no universal analytical solution exists that is suitable for all hydrogen production projects. This is due to the considerable variation in system parameters depending on the technology, scope and application. However, the basic principles of sampling, preparation and analysis are largely comparable across the hydrogen industry.

Moving forward, it is essential that research and development initiatives prioritise additional standardisation, enhanced digital integration and reinforced robustness of analysis systems in dynamic operating conditions. Reliable and precise process analytics should ultimately be regarded as a fundamental technology that helps the economic and sustainable growth of green hydrogen production.

References

1. M. G. Buonomenna, A. Patonia, Green Hydrogen in Europe: Where Are We Now?, Hydrogen, 6,105 (2025). https://doi.org/10.3390/hydrogen6040105
2. T. Reisinger, Principles and concepts of safety and quality measurement in H_2 production, ACHEMA 2024, Frankfurt a. M., June 10-14 (2024).
3. H. Mahler, F. Thielland, Process analytics secures the quality of green hydrogen and CO_2 pipeline management, ptc 2023, Berlin, May 8-11 (2023).
4. J. Erens, T. Reisinger, Probe Sampling in Hazardous Areas under Extreme Conditions, CEM Middle East 2025, Bahrain, April 8-10 (2025).
5. H. Mahler, Energy-efficient and sustainable over the entire life cycle, Envirotech-online.com | IET September / October 38-39 (2025). https://www.envirotech-online.com/news/air-monitoring/6?page=6
6. J. Gorvett (Ed.), in IRENA, Green Hydrogen Cost Reduction 35 (2020). www.irena.org/publications

 t pp s/dl o.gpr5dl 0gl 0j 1de: 3conf(606) (j 60100(

Reducing helium dependency in gas chromatography for natural gas blends with hydrogen

Saskia Kunkeler [1*], *Eva* Krolis[1], and *Füsun* Karaburun-Yavuz[1]

[1]Qmicro B.V., Neptunusstraat 21, 7521 WC, Enschede, The Netherlands

Abstract. The transition toward blending hydrogen in natural gas networks requires a new generation of gas chromatographs, capable of accurately measuring hydrogen. At the same time, the ongoing helium shortage highlights the need to reduce helium consumption in gas chromatography. This research evaluates the analytical performance of the DynamiQ-X NG2220 micro gas chromatograph using three carrier gas combinations: helium-argon, hydrogen-argon and argon alone. Ten natural gas blends containing up to 20% hydrogen were measured to assess calorific value accuracy, limits of detection, and overall performance in accordance with OIML R140 and ISO 10723. All carrier gas combinations produced calorific values within the limits of OIML R140 accuracy class A. Operation with argon as the only carrier gas showed lower detector response and therefore higher limits of detection but remained suitable for fiscal metering ranges. Carrier gas consumption was low for all methods, and replacing helium with a hydrogen or argon can save approximately 19.2 litres of helium per day during continuous operation. Overall, the DynamiQ-X NG2220 demonstrated reliable performance with all tested carrier gases without needing any hardware changes, supporting the use of hydrogen or argon as viable alternatives to helium for online monitoring of natural gas blends with hydrogen.

1 Introduction

Injecting sustainably produced hydrogen into the natural gas grid can significantly reduce greenhouse gas emissions. Many European governments have implemented policies to support hydrogen blending of up to 20% in their existing natural gas networks [1, 2]. This shift presents challenges for composition control in fiscal metering, as currently installed gas chromatographs equipped with thermal conductivity detectors (TCDs) cannot measure hydrogen without an additional column and detector, as well as an additional carrier gas [1]. Since helium generally cannot be used for measuring hydrogen due to similar thermal conductivities, argon or nitrogen are required.

At the same time, the ongoing helium shortage further emphasizes the need to reduce helium consumption in gas chromatography [3]. As helium becomes increasingly scarce and expensive, instruments that operate reliably with alternative carrier gases, or that minimize carrier gas usage, will gain importance. Hydrogen, with its high thermal conductivity, can serve as a fitting substitute for helium, but its use introduces safety concerns. Hydrogen is highly flammable when mixed with air, has low ignition energy and permeates materials easily, making leaks likely to ignite. For this reason, hydrogen carrier gas is often generated onsite using a generator, though this is not always feasible or permitted under local legislation and safety guidelines.

Additionally, for the measurement of hydrogen in natural gas blends, argon or nitrogen remain essential. Switching exclusively to argon can address this, but its significantly lower thermal conductivity results in reduced signal response which may impact measurement accuracy or limits of detection.

The DynamiQ micro Gas Chromatograph (GC) used in this study incorporates several MEMS micro-machined chip components, connected through a patented chip to chip technology. This architecture enables a compact instrument footprint in an explosion proof design [4]. The DynamiQ-X NG2220 version was developed for online measurements of natural gas blends with hydrogen, providing a full analysis within 45 seconds. It is certified to meet OIML R140 accuracy class A requirements [5, 6], using helium and argon as carrier gases.

In this study, the DynamiQ-X NG2220 was evaluated using three carrier gas combinations, helium and argon, hydrogen and argon and solely argon, across a range of natural gas compositions containing hydrogen. No hardware changes were required when switching between carrier gases. The accuracy of the derived calorific values was assessed by comparison to the international recommendation OIML R140 [5]. Additionally, performance evaluations following ISO 10723 [7] were conducted to assess measurement errors and associated uncertainties. Limits of detections for all individual components were determined, and the carrier gas consumption was recorded. Together, these evaluations provide a basis for assessing the suitability of different carrier gas combinations for fiscal metering of natural gas blends with hydrogen.

* Corresponding author: saskia.kunkeler@qmicro.com

2 Materials and methods

2.1 Equipment

The DynamiQ-X NG2220 contains three GC units that each perform in parallel a different GC analysis under individually optimized conditions using thermal conductivity detectors (TCD). The instrument consists of a stream selector with 4 sample streams. One sample stream can be measured at the same time but the three GC units analyze this gas simultaneously with parameters optimized for different components, as depicted in figure 1.

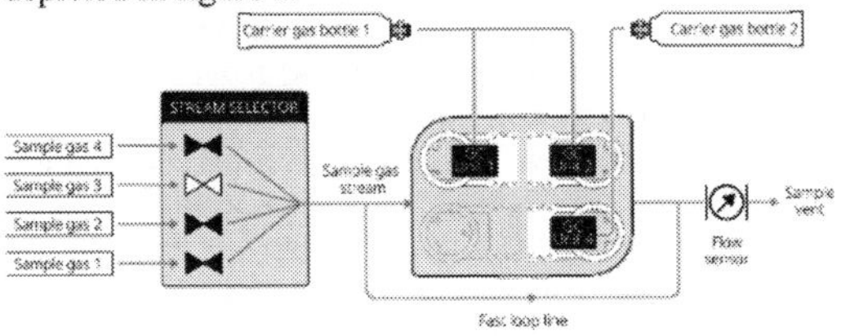

Fig. 1. Schematic of the stream selector and GC units.

The three GC units consist of an injector and two detectors integrated on a chip, a pre-column and an analytical column. This configuration enables the use of backflush technology, which decreases the analysis time and prevents late-eluting or unwanted compounds from entering the analytical column. During sample injection, the system operates in foreflush mode. At a defined point in the analysis, the flow through the pre-column is reversed and directed to the backflush detector (TCD BF), where a combined peak of all late-eluting components is recorded. Simultaneously, components that have already reached the analytical column continue to elute at the foreflush detector (TCD FF). A schematic is given in Figure 2.

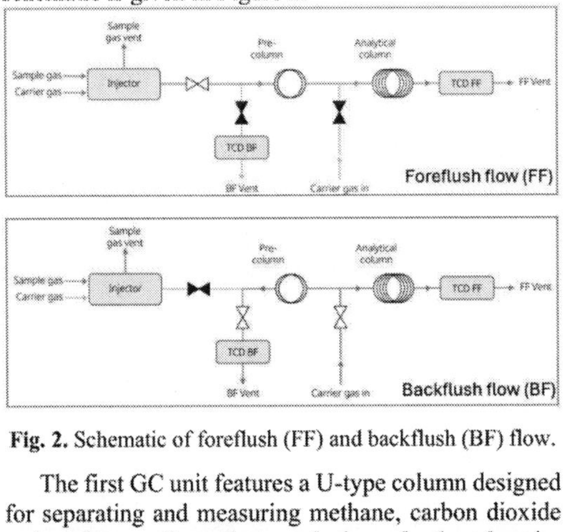

Fig. 2. Schematic of foreflush (FF) and backflush (BF) flow.

The first GC unit features a U-type column designed for separating and measuring methane, carbon dioxide and ethane. For the analysis of the heavier hydrocarbons, the second GC unit is used, equipped with a P-type column. This column effectively separates propane, butanes and pentanes. All hexanes and higher hydrocarbons are directed to the backflush detector, resulting in a combined peak. The DynamiQ-X NG2220 includes a third GC unit to measure permanent gases using an M-type column. This column is used to measure hydrogen, oxygen and nitrogen [6].

The third GC unit can be used with a different carrier gas than GC units 1 and 2, as depicted in Figure 1. For this study, the carrier gas for GC units 1 and 2 was varied between helium, hydrogen and argon, while the carrier gas for GC unit 3 was argon for all measurements.

The DynamiQ-X is designed for continuous monitoring and works therefore stand-alone. The generated data can be communicated via Modbus and the instrument is WELMEC 7.2 compliant [6, 8]. The ATEX/IECEx certified analyzer is housed in an explosion-safe marine-grade aluminium IP65 enclosure. The instrument has a volume of 10 L and a low carrier gas consumption of one gas bottle for over one year [4].

The thermal conductivity detectors (TCDs) operate by detecting changes in heat transfer from a heated filament as different components pass through. The magnitude of the signal is determined by the difference in thermal conductivity between the carrier gas and the analyte. Helium and hydrogen are commonly used carrier gases because of their high thermal conductivities. In contrast, argon has a lower thermal conductivity, leading to a lower signal response.

2.2 Sample gases

A set of ten Certified Reference Materials (CRMs), containing 13 components, was used in this study. The ranges were chosen to be representative of, or broader than, those encountered in fiscal metering of natural gas with added hydrogen up to 20%. Table 1 summarizes the concentration ranges, with the mixtures distributed approximately evenly across these ranges. The sample gases were prepared by high precision gravimetry in accordance with ISO 6142-1 [9] and calibrated by traceable comparison methods compliant with ISO 6143 [10] by the ISO 17025-accredited laboratory EffecTech [11].

Table 1. Component ranges.

Component	Range sample gases
Methane	65-93%
Carbon dioxide	0.1-12%
Ethane	0.1-15%
Propane	0.1-6%
2-methylpropane	0.01-1.2%
n-Butane	0.01-1.2%
2,2-dimethylpropane	0.01-0.06%
2-methylbutane	0.01-0.5%
n-Pentane	0.01-0.5%
n-Hexane (C6+)	0.01-0.25%
Hydrogen	0.01-20%
Oxygen	0.01-3%
Nitrogen	0.1-20%

t pp s/dd o.gpr5dl 0gl 0j 1de: 3confd606) (j 60100(

2.3 Methodology

The DynamiQ-X NG2220 was factory-tuned and calibrated for operation with helium and argon carrier gases, for which the standard analysis time is 45 seconds. For the hydrogen-argon carrier gas combination, the carrier pressure and the backflush time of GC units 1 and 2 were reduced to account for the lower viscosity of hydrogen. In contrast, argon has a higher viscosity and a lower thermal conductivity than both helium and hydrogen. Consequently, when argon was used as the only carrier gas, the carrier pressure and backflush time were increased to shorten the retention times and sharpen the peaks. Additionally, the injection time was increased to compensate for the reduced detector response associated with argon. After calibration, each CRM was measured 10 times for every carrier gas combination.

All measurement methods were stored in the instrument's internal memory, and no hardware changes were required when switching between carrier gases. The only waiting time involved was allowing the detectors to stabilize, enabling the instrument to resume operation within 30 minutes after changing carrier gases.

3 Results and discussion

3.1 Calorific value

The calorific value (CV) was determined following ISO 6976 [12] using a combustion temperature of 15 °C, a metering temperature of 15 °C and a metering pressure of 101.325 kPa. The resulting relative errors for each carrier gas combination are given in Table 2. These errors were evaluated against the limits specified in OIML R140, which defines performance requirements for measuring systems of gaseous fuel [5]. For accuracy class A, the maximum permissible relative error is 0.50% and all CVs fall within this limit.

Table 2. Relative error on the calorific value.

Sample gas	Actual CV (MJ/m^3)	Error (%)		
		Helium Argon	Hydrogen Argon	Argon Argon
Gas 1	25.92	0.09	-0.41	0.16
Gas 2	31.79	0.10	-0.02	-0.02
Gas 3	34.01	0.16	0.11	-0.07
Gas 4	39.29	-0.01	-0.02	0.02
Gas 5	39.31	-0.04	-0.05	0.06
Gas 6	45.47	-0.01	-0.12	-0.07
Gas 7	26.12	0.28	-0.46	0.30
Gas 8	35.88	-0.07	-0.10	-0.01
Gas 9	40.21	-0.44	-0.35	-0.44
Gas 10	45.23	-0.03	-0.14	0.06

The largest deviations occur at the extremes of the CV range, which coincide with mixtures containing the highest concentrations of some components. Nevertheless, all errors remain well within the accuracy class, and there are no substantial differences found between the different carrier gas configurations.

3.2 Performance evaluation – ISO 10723

ISO 10723 [7] describes a procedure for a performance evaluation of a natural gas analyzer to determine the errors and associated uncertainties on parameters such as the calorific value. In accordance with this method, the instrument response was first characterized using the ten CRMs. A Monte-Carlo simulation was then performed to construct a data set of 10.000 hypothetical, naturally occurring compositions of natural gas. For each simulated composition, the error was obtained by taking the difference between the calorific value determined from the true and measured composition.

Following the nomenclature in ISO 10723, the absolute value of the mean error in calorific value ($|\overline{\delta CV}|$) and the sum of this error with its expanded uncertainty ($|\overline{\delta CV}| + U_c(\overline{\delta CV})$) were determined for all three carrier gas combinations. The former can be compared to a pre-defined maximum permissible bias (MPB), while the latter can be evaluated against a maximum permissible error (MPE), depending on the application requirements.

Table 3 depicts the calculated errors and associated uncertainties. The error in calorific value for the helium-argon and hydrogen-argon carrier gas combinations are comparable. When argon is used as the sole carrier gas, the errors increase, primarily due to the lower detector response for low concentration components, particularly the C6+ fraction. However, the expanded uncertainties of all three carrier gas combinations are of similar magnitude. As a result, the sum of the error with its expanded uncertainty, which is used to assess compliance with a maximum permissible error, shows a much smaller difference between the three carrier gas configurations.

Table 3. Errors and uncertainties from ISO 10723 performance evaluation.

	Helium Argon	Hydrogen Argon	Argon Argon
$|\overline{\delta CV}|$ (MJ/m^3)	0.0019	0.0022	0.0287
$|\overline{\delta CV}| + U_c(\overline{\delta CV})$ (MJ/m^3)	0.0371	0.0637	0.0853

3.3 Limits of detection

Due to differences in thermal conductivity between helium, hydrogen and argon, the detector response, and therefore the peak size, varies with the carrier gas. The limit of detection (LOD) per component is calculated as three times the standard deviation of the noise, converted to concentration:

$$LOD = \frac{3 \times Noise}{RF} \quad (1)$$

where *Noise* is the RMS noise and *RF* is the response factor of the impurity. The response factor is obtained from the calibration curve, by dividing the measured peak area by the analyte concentration.

When argon is used as the sole carrier gas, the noise level is slightly higher. Additionally, hydrocarbons show lower response factors compared to helium or hydrogen because the difference in thermal conductivity is smaller. Together, these effects negatively impact the LOD for measurements, resulting in higher values. Hydrogen behaves similarly to helium in both noise level and response factor, which explains why the LODs for hydrogen-argon and helium-argon measurements are comparable. Table 4 summarizes the limits of detection for all components per carrier gas combination used.

Table 4. Limits of detection (LOD) per component.

Component	Limit of detection (ppm)		
	Helium Argon	**Hydrogen Argon**	**Argon Argon**
Methane	0.3	0.2	1.1
Carbon dioxide	0.2	0.2	5.1
Ethane	0.4	0.3	2.7
Propane	0.2	0.1	0.5
2-methylpropane	0.2	0.1	0.6
n-Butane	0.2	0.1	0.7
2,2-dimethylpropane	0.3	0.2	1.2
2-methylbutane	0.3	0.2	1.3
n-Pentane	0.3	0.2	1.7
C6+	0.3	0.2	16.2
Hydrogen	0.2	0.2	0.2
Oxygen	1.6	1.6	1.5
Nitrogen	3.2	3.0	3.1

3.4 Carrier gas usage

The carrier gas consumption remains roughly the same across the different carrier gas combinations. Thanks to the short cycle time of only 45 seconds, the total consumption is about 15 ml per measurement. Switching from helium-argon to either hydrogen-argon or argon-argon carrier gas eliminates the use of helium, saving approximately 10 ml of helium per measurement. In fiscal metering, continuously measuring is common, corresponding to 1920 injections per day. Under these conditions, replacing helium-argon with an alternative carrier gas combination would save approximately 19.2 litres of helium per day.

4 Conclusion

This study assessed the DynamiQ-X NG2220 micro GC for fiscal metering of natural gas blends with hydrogen using helium-argon, hydrogen-argon and argon-only carrier gas combinations. All three setups produced calorific values within the limits of OIML R140 accuracy class A. ISO 10723 performance evaluations showed that helium-argon and hydrogen-argon carrier gases deliver nearly identical errors and uncertainties. When argon is used as the sole carrier gas, the errors increase, primarily due to the lower detector response for low concentration components. However, the combined error and expanded uncertainty remain comparable and suitable for practical use.

The analysis of the limits of detection showed that helium and hydrogen provide the strongest detector response, whereas argon yields higher LODs. Despite this, all LODs were sufficient for the concentration ranges relevant to fiscal metering. The carrier gas consumption was low for all methods and switching from helium-argon to either hydrogen-argon or argon-argon can save roughly 19.2 litres of helium per day under continuous operation.

Overall, the instrument functions reliably with all tested carrier gases without any hardware changes needed. Hydrogen provides a strong alternative to helium where safety regulations permit. Using argon as the only carrier gas is feasible with some loss in sensitivity but significant saving of helium consumption and with the benefit that only one carrier gas bottle is required.

References

1. Fraunhofer Institute for Energy Economics and Energy System Technology, The limitations of hydrogen blending in the European gas grid (2022)
2. K. Kanellopoulos, S. Busch, M. De Felice, S. Giaccaria, Blending hydrogen from electrolysis into the European gas grid. Publications Office of the European Union **JRC126763** (2022)
3. D. Kramer, Helium prices surge to record levels as shortage continues. Physics Today **76(9),** 18-20 (2023)
4. Qmicro by Sensirion, DynamiQ-X process GC – Gas chromatographs for on-line gas analysis (2024)
5. International Organization of Legal Metrology, International Recommendation OIML R140 – Measuring systems for gaseous fuel (2007)
6. Qmicro by Sensirion, Application note – Natural gas and natural gas blended hydrogen (2025)
7. International Organization for Standardization (ISO), ISO 10723:2012 Natural gas – Performance evaluation for analytical systems (2019)
8. European Cooperation in Legal Metrology (WELMEC) e.V., WELMEC Guide 7.2 –

Software Guide (EU Measuring Instruments Directive 2014/32/EU) – Version 2022 (2022)
9. International Organization for Standardization (ISO), ISO 6142-1:2015 Gas Analysis – Preparation of calibration gas mixtures – Part 1: Gravimetric method for Class I mixtures (2015)
10. International Organization for Standardization (ISO), ISO 6143:2001 Gas analysis – Comparison methods for determining and checking the composition of calibration gas mixtures (2001)
11. International Organization for Standardization (ISO), ISO 17025:2017 General requirements for the competence of testing and calibration laboratories (2017)
12. International Organization for Standardization (ISO), ISO 6976:2016 Natural gas – Calculation of calorific values, density, relative density and Wobbe indices from composition (2016)

EPJ Web of Conferences **352**, 01004 (2026) https://doi.org/10.1051/epjconf/202635201004
GAS 2026

Gas monitoring systems and calorimeters for hydrogen-containing natural gas applications

Ken Uchiyamada and *Kenichi* Kojima*

RIKEN KEIKI Co., Ltd., Research and Development Department, 2-3, Minamisakae-cho, Kasukabe-shi, Saitama, Japan

Abstract. This study proposes semi-custom-made gas monitoring systems that combine our existing gas detection alarms and calorimeters to support decarbonization technologies. These systems require no catalytic combustion or separate columns, and the use of physical sensors eliminates the need for frequent cleaning or recalibration. Even with preprocessing, the unit remains compact and ensures straightforward maintenance. Application examples include monitoring hydrogen-containing natural gas, analyzing ammonia synthesis and decomposition, evaluating methanation, and assessing steel-mill by-product gases.

1 Background

The global trend toward decarbonization to mitigate climate change is accelerating, and green energy is gaining significant attention. Among the various green energy sources, hydrogen has emerged as one of the most promising options, and its utilization is currently being demonstrated worldwide. With this shift, the demand for gas calorimetry is increasing. While several calorimeters are available for natural gas, there are very few devices capable of measuring hydrogen-containing natural gas in real-time with high accuracy.

2 Calorimeter principle

We developed the calorific value determining device OHC-800, as shown in Figure 1. The calorific value of a gas can be determined using the correlation among its speed of sound (measured by a sonic sensor), its refractive index (measured by an optical sensor), and the calorific value itself.

Fig. 1. OHC-800 calorimeter.

For gases composed mainly of paraffinic hydrocarbons, each sensor can accurately measure calorific value. However, when the gas contains components with low calorific value (e.g., CO, N_2, CO_2), these introduce measurement errors. By combining the signals from both sensors, these errors can be canceled.

2.1 Sonic sensor

The sonic sensor determines the speed of sound through precise time-of-flight measurements, using a pair of piezoelectric elements—one for transmission and one for reception—positioned at opposite ends of the gas cell, as shown in Figure 2.

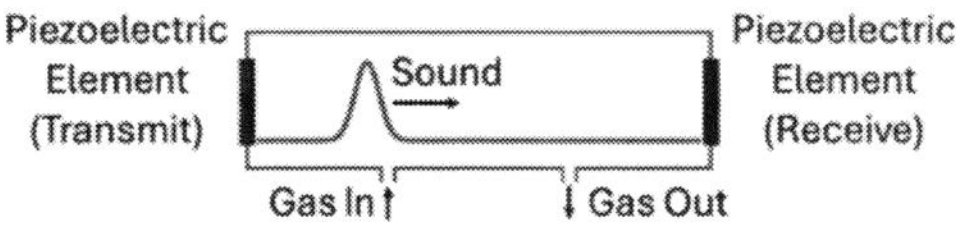

Fig. 2. Sonic sensor image.

Figure 3 illustrates the relationship between the speed of sound and the calorific value of various gases. The straight line connecting the ■ marks represents the function Q_{sonic}, which describes the correlation between calorific value and speed of sound in mixtures composed of paraffinic hydrocarbons.

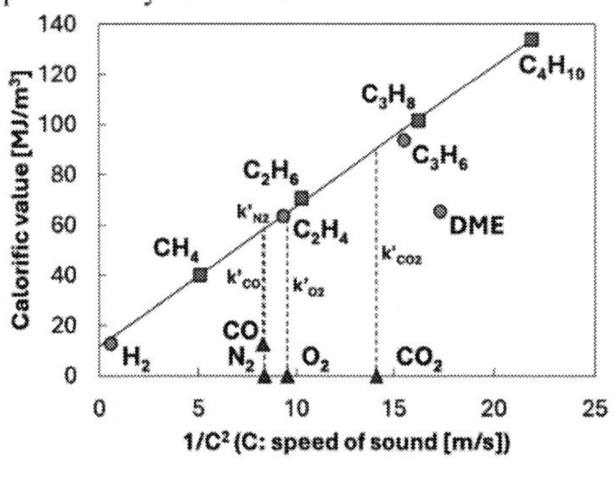

Fig. 3. Speed of sound vs. calorific value.

* Corresponding author: k-kojima@rikenkeiki.co.jp

2.2 Optical sensor

The optical sensor measures the refractive index by detecting interferometric fringe shifts. The device comprises two gas chambers: one filled with a reference gas—typically air or N_2—and the other with the measurement gas, as shown in Figure 4.

Light emitted from an LED is split into two paths, passes through both chambers, and is then recombined before reaching the camera. The phase shift arising from the difference between the two gases generates an interferometric fringe shift, which is captured by the camera.

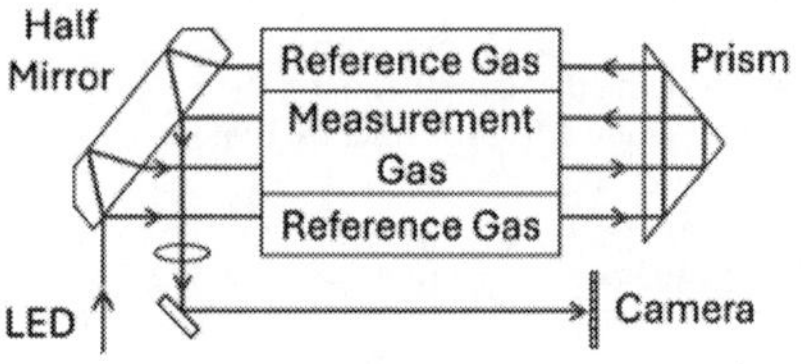

Fig. 4. Optical sensor image.

Figure 5 illustrates the relationship between refractive index and calorific value of various gases. The straight line connecting the ■ marks represents the function Q_{opt}, which describes the correlation between calorific value and refractive index in mixtures composed of paraffinic hydrocarbons.

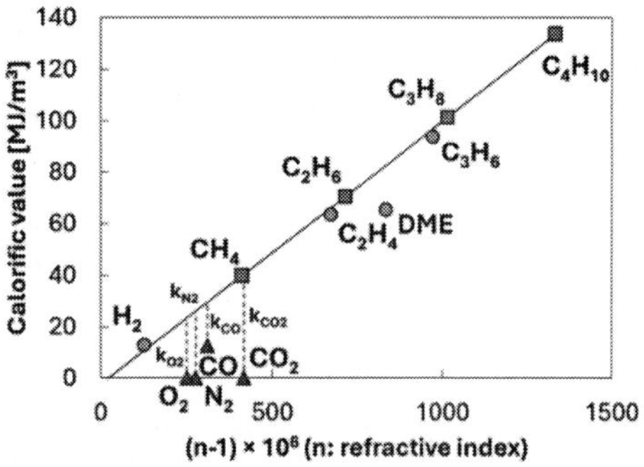

Fig. 5. Refractive index vs. calorific value.

2.3 Opt-sonic calculation

The ratio of calorific values measured by the sonic sensor and the optical sensor remains approximately constant for most interfering gases (e.g., CO, N_2, CO_2), as summarized in Table 1. This characteristic allows the measurement errors introduced by these interfering gases to be effectively canceled when the outputs of both sensors are combined.

Table 1. Error factor of two sensors.

Gas	Optical Sensor	Sonic Sensor	Ratio
CO_2	k_{CO2} = 40.52 MJ	k'_{CO2} = 89.84 MJ	2.21
CO	k_{CO} = 17.14 MJ	k'_{CO} = 45.24 MJ	2.63
N_2	k_{N2} = 26.19 MJ	k'_{N2} = 58.18 MJ	2.22
O_2	k_{O2} = 23.60 MJ	k'_{O2} = 64.56 MJ	2.75
DME	k_{DME} = 17.32 MJ	k'_{DME} = 42.43 MJ	2.45

Let the true calorific value of a natural gas, including interfering gases, be represented by Q. Then:

$$Q = Q_{opt} - (k_{CO2} \cdot x_{CO2} + k_{CO} \cdot x_{CO} + k_{N2} \cdot x_{N2} + k_{O2} \cdot x_{O2}) \quad (1)$$
$$Q = Q_{sonic} - (k'_{CO2} \cdot x_{CO2} + k'_{CO} \cdot x_{CO} + k'_{N2} \cdot x_{N2} + k'_{O2} \cdot x_{O2}) \quad (2)$$

where:

· x_i denotes the volume fraction of each interfering gas component.

· k_i and k'_i represent the error coefficients corresponding to the influence of each component.

Although these coefficients differ for each gas, their ratio—denoted as α—remains nearly constant:

$$k_{CO2} \approx \alpha \cdot k'_{CO2},\ k_{CO} \approx \alpha \cdot k'_{CO},\ k_{N2} \approx \alpha \cdot k'_{N2},\ k_{O2} \approx \alpha \cdot k'_{O2} \quad (3)$$

Using equation (3), the opt-sonic calculation formula is obtained:

$$Q = Q_{opt} - \frac{(Q_{opt} - Q_{sonic})}{1-\alpha} \quad (4)$$

Figure 6 illustrates the correlation between calorific values calculated from equation (4) and the actual values. All data points align along a straight line with a slope of 1, confirming that accurate calorific values can be obtained—even when the natural gas contains interfering components—by combining Q_{opt} and Q_{sonic}.

The same approach can also be applied to determine the specific gravity of natural gas, providing a unified method for characterizing gas properties.

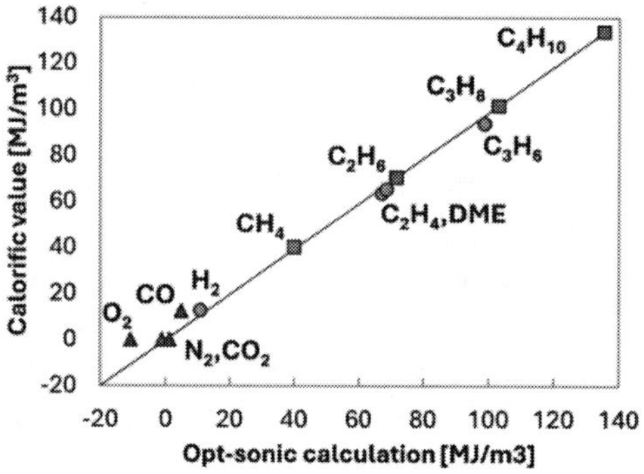

Fig. 6. Opt-sonic calculation results.

3 H2-containing natural gas application

The opt-sonic calculation method is applicable not only to natural gas but also to hydrogen-containing natural gas mixtures. We propose several approaches for determining both the calorific value and the hydrogen concentration within such blended gases.

Some of these approaches utilize techniques that combine multiple gas sensors with a programmable logic controller (PLC). By leveraging our extensive sensor lineup, it is possible to measure multiple parameters of mixed gases by integrating different sensors into a unified system. We refer to this integrated solution as the Real-Time Gas Monitoring System (RTGMS).

3.1 Measurement of calorific value

The OHC-800 calorimeter was certified as OIML R140 Class A in 2023. The instrument is applicable to both natural gas and hydrogen-containing natural gas, covering a calorific value range of 30.1–47.3 MJ/m³. The certified working ranges for each gas component are summarized in Table 2.

Table 2. Working range of OIML-certified OHC-800.

Component	Working range [mol%]	
	Natural gas	Natural gas + H_2
CH_4	70–100	67–100
C_2H_6	0–15	0–15
C_3H_8	0–5	0–9
C_4^+	C_4^+ shall be converted to the equivalent CH_4 + C_3H_8 mixture; the result must fall within their concentration ranges.	
N_2	0–20	0–10
CO_2	0–5	0–6
H_2	0	0–20
CV [MJ/m³]	30–47.5	

The OHC-800 enables continuous calorific value measurement of natural gas and mixtures containing up to 20 mol% hydrogen. The system provides a measurement-data update interval of 0.25 s, a 90% response time below 5 s, and a measurement uncertainty within ±0.5%, as shown in Figure 7. These performance characteristics exceed those achievable with conventional calorimeters and gas chromatographs, particularly in applications requiring rapid response and real-time monitoring.

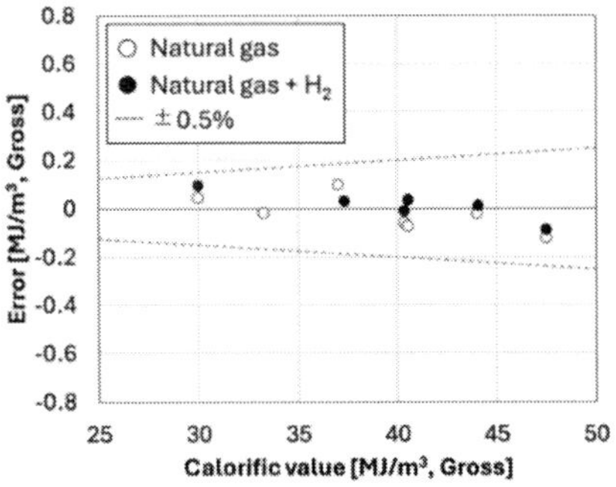

Fig. 7. Measurement results of OIML-certified OHC-800.

3.2 Measurement of added H_2 proportion

If the measurement of natural gas prior to hydrogen addition is possible, the RTGMS extends the effective working range for hydrogen in calorific value measurement up to 100 vol% and enables the proportion of added hydrogen to be measured.

In this method, calorific value and refractive index of the original natural gas are first measured using the OHC-800 before hydrogen is added. After hydrogen addition, the refractive index of the resulting mixture is measured using the FI-900, which is equipped solely with an optical sensor. The calorific value and hydrogen concentration of the mixture are then calculated by a PLC, as shown in Figure 8. The specific gravity of the mixture can also be determined using the same approach.

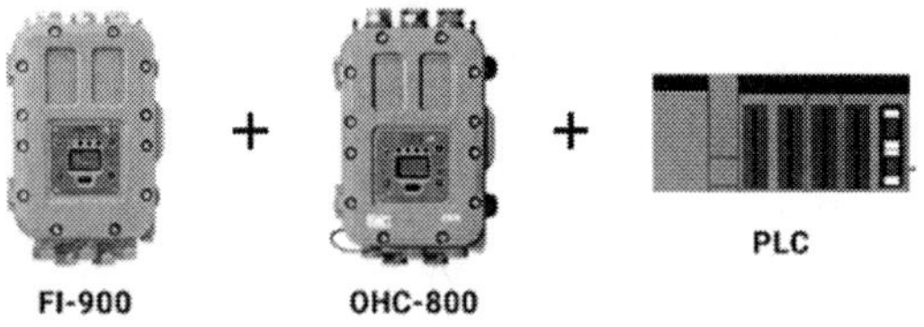

Fig.8. RTGMS for H_2 + natural gas measurement.

Table 3 presents the measurement and calculation results for hydrogen concentration, calorific value, and specific gravity when hydrogen was incrementally added to natural gas using this system. The theoretical hydrogen addition rate corresponds to the ideal volumetric mixing ratio. The results show that the calculated values closely match the theoretical values, demonstrating that the system provides highly consistent and reliable measurement.

Table 3. Measurement results for H_2 added natural gas.

Added H_2 [%]		CV [MJ/m³]		Specific Gravity	
Theo.	Meas.	Theo.	Meas.	Theo.	Meas.
0	0.0	45.40	45.33	0.647	0.645
10	9.3	42.13	42.27	0.589	0.591
20	19.1	38.86	39.08	0.531	0.534
30	29.2	35.60	35.81	0.473	0.476
40	39.4	32.34	32.49	0.416	0.417
50	49.5	29.08	29.20	0.358	0.359
60	60.1	25.82	25.76	0.300	0.298
70	70.2	22.56	22.48	0.243	0.241
80	80.2	19.31	19.20	0.185	0.181
90	89.9	16.05	16.04	0.127	0.127
100	99.5	12.79	12.94	0.070	0.072

3.3 Measurement of H_2 concentration

The RTGMS—consisting of thermal conductivity type hydrogen sensors, the OHC-800, and a PLC—enables an extension of the measurable range to 60 vol% hydrogen for calorific value determination and up to 100 vol% for hydrogen concentration measurement. In this method, the difference in output from the thermal conductivity sensor at two distinct sensor temperatures is utilized. As shown in Figure 9, the sensor output remains essentially zero for the major components of natural gas, except for N_2 and CO_2.

The influence of N_2 and CO_2 can be corrected using the OHC-800, as expressed in equation (5):

$$x_{H2} = \Delta I - a \times (Q_{opt} - Q_{sonic}) \quad (5)$$

where:

· x_{H2} denotes the volume fraction of hydrogen gas.

· ΔI represents the difference in thermal conductivity sensor output at two sensor temperatures.

· a is a constant representing the correction factor for N_2 and CO_2 induced errors.

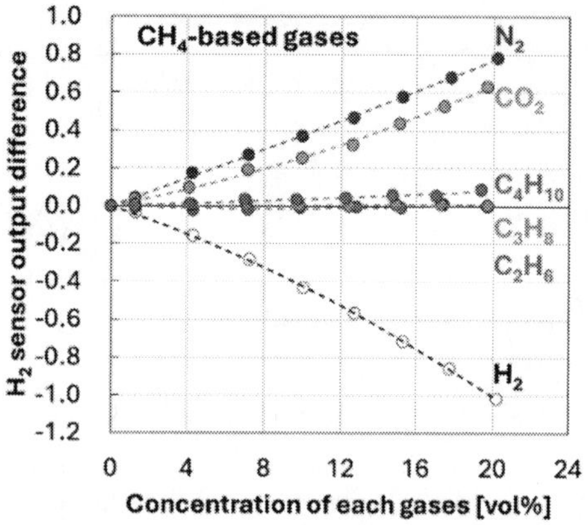

Fig. 9. H_2 sensor output vs. gas concentration.

Measurements were conducted using several natural gas compositions, as summarized in Table 4. Hydrogen was added to each natural gas sample in the range of 0 to 100 vol%, and both the hydrogen concentration and calorific value were evaluated. As shown in Figure 10, the system achieved measurement errors within ±0.5% for calorific value and within ±1.0 vol% for hydrogen concentration, demonstrating high measurement accuracy across the entire hydrogen addition range.

Table 4. Composition of measured natural gas.

Composition [vol%]						CV
CH_4	C_2H_6	C_3H_8	i-C_4H_{10}	N_2	CO_2	[MJ/m³]
80	10	2	1	5	2	42.32
90	2	1	0	5	2	38.35
90	4	2	0	3	1	40.77
83	10	3	0	3	1	43.19
90	4	2	1	3	0	42.10
82	10	4	1	3	0	45.14
90	6	3	1	0	0	44.52
85	10	5	0	0	0	46.33

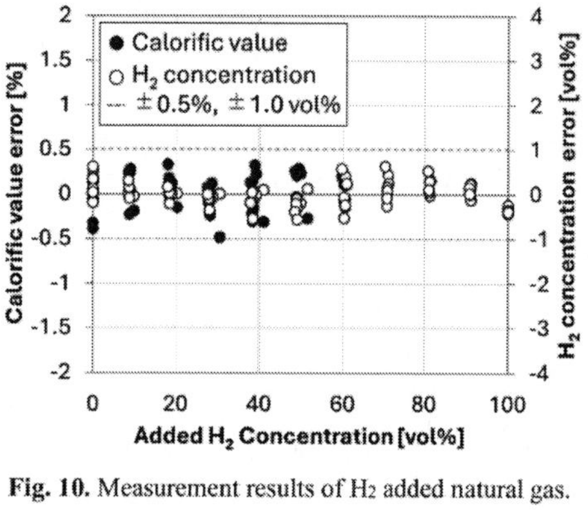

Fig. 10. Measurement results of H_2 added natural gas.

4 Other applications

4.1 Ammonia synthesis and decomposition

Ammonia is considered an essential energy source for carbon neutrality, contributing to decarbonization across various sectors. As a result, there is a growing demand to accelerate its rapid social implementation. Compared with the direct transport of hydrogen, ammonia offers significant advantages in terms of safety, storage density, and feasibility for large-scale, long-distance transportation. For these reasons, technologies for ammonia decomposition systems have been developed by several research groups, under the assumption that ammonia will be transported to ports near industrial facilities and subsequently decomposed into H_2 and N_2.

The RTGMS, consisting of the OHC-800 and a PLC, enables continuous monitoring of gas composition during ammonia synthesis and decomposition processes. Gas mixtures composed of NH_3, H_2, and N_2 exhibit a triangular correlation between refractive index and specific gravity, as shown in Figure 11. Using the refractive index and specific gravity coordinates from the OHC-800, the PLC calculates the concentrations of NH_3, H_2, and N_2 in real time.

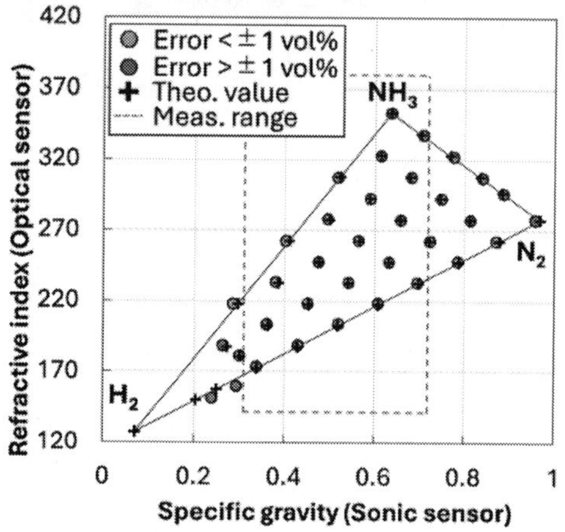

Fig. 11. Refractive index vs. specific gravity.

Figure 12 illustrates the measurement performance of the system for NH_3, H_2, and N_2 mixtures with a specific gravity range of 0.31–0.72, achieving an accuracy of ±1.0% under calibrated conditions. For applications requiring higher N_2 concentrations, an alternative configuration supports a specific gravity range of 0.43–0.97. Multiple RTGMS units have already been deployed in operational ammonia synthesis and decomposition facilities, demonstrating the system's applicability and stability under real-world process conditions.

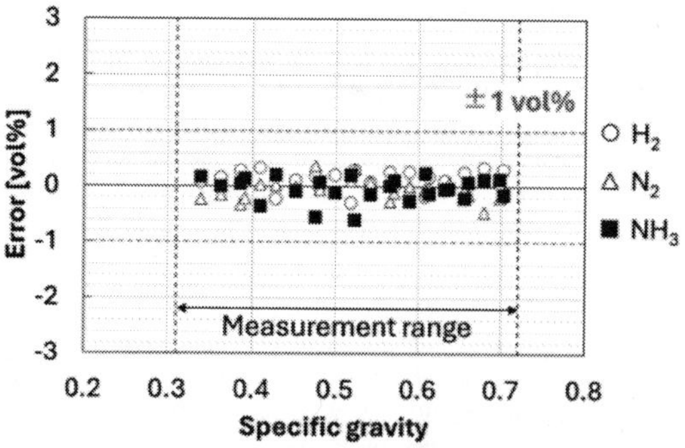

Fig. 12. Measurement results of NH_3, H_2, N_2 concentrations.

4.2 Methanation

Methanation is a process in which CH_4 is synthesized from H_2 and CO_2, and it is increasingly recognized as a key technology for producing carbon-neutral city gas. Using the same analytical approach described in the previous section for ammonia synthesis and decomposition analysis, the RTGMS, consisting of the OHC-800 and a PLC, enables real-time monitoring of CH_4, H_2, and CO_2 concentrations within a specific gravity range of 0.31–0.72, as well as the calorific value, throughout the methanation reaction.

Figure 13 illustrates an example of system performance, showing the behavior of calorific value and calculated gas composition when the mixture was alternately varied from pure methane (CH_4 99.9995 vol%) to a mixture of CH_4: 90 vol%, H_2: 5 vol%, and CO_2: 5 vol%.

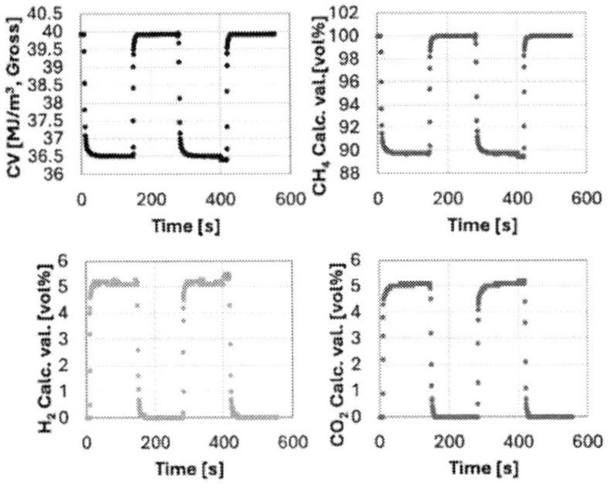

Fig. 13. Methanation gas measurement results.

The calculated results for each composition remained within an error margin of approximately ±2%, and the system exhibited a 90% response time of roughly 20 s. As shown in Figure 14, multiple RTGMS units have already been deployed in operational methanation facilities, as well as in the ammonia synthesis and decomposition facilities described in the previous section.

Fig. 14. OHC-800 deployed in a methanation plant.

4.3 Steel-mill by-product gas

The Japanese steel industry is recognized for achieving the highest level of energy efficiency worldwide in both converter and electric-furnace steelmaking, with comprehensive energy-saving measures already in place.

In February 2021, the *Basic Policy for the Japanese Steel Industry Regarding Japan's Carbon Neutrality by 2050* was formulated, outlining strategies to further reduce CO_2 emissions across the entire supply chain. These strategies emphasize two primary value propositions:

1. the provision of high-performance steel products and solutions that contribute to CO_2 reduction across society, and
2. the realization of carbon-neutral steel through decarbonization of the steelmaking process.

To monitor the composition of steelmaking by-product gases, the RTGMS—comprising the OHC-800, CO and CO_2 infrared detectors (both from our product lineup), and a PLC—is employed. This system enables real-time calculation of gas compositions containing CO, CO_2, CH_4, H_2, and N_2.

Fig. 15. CO and CO_2 detectors.

Figure 16 shows the relationship between the calculated and theoretical concentrations of each component in mixture gases prepared by blending the cylinder gases listed in Table 5 at various mixing ratios. These cylinder gases simulate typical COG (coke oven gas), BFG (blast furnace gas), and LDG (converter furnace gas). The theoretical values represent ideal volumetric mixing ratios, and the calculated composition values exhibit errors within ±2%, demonstrating the high accuracy of the system.

Table 5. Composition of cylinder gases.

	Composition [mol%]				
	CO	CO_2	CH_4	H_2	N_2
COG	7.5	3.5	27.3	56.4	5.3
BFG	22.0	20.4	0.0	2.5	55.1
LDG	58.5	18.5	0.0	1.5	21.5

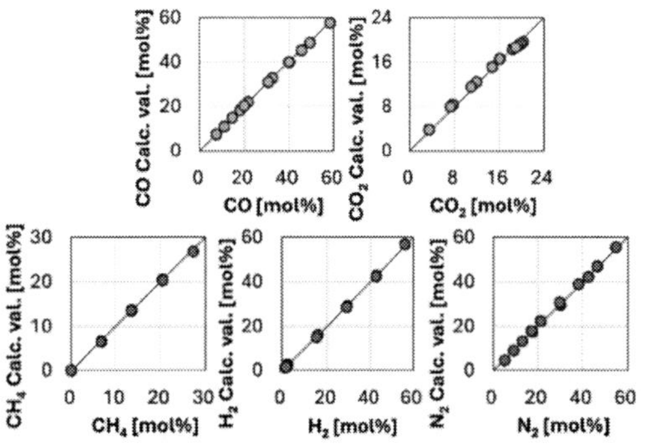

Fig. 16. Measurement results of steel-mill by-product gases.

5 Conclusion

The OHC-800's dual-sensor opt-sonic calculation method provides a robust and maintenance-friendly approach for precise and rapid calorific value determination in complex gas streams. Its capability for continuous, real-time measurement supports a wide range of industrial operations and enables the reliable utilization of hydrogen-containing natural gas. These features facilitate timely decision-making in the energy, chemical, and environmental monitoring sectors, contributing to safer and more efficient industrial processes.

Furthermore, our company has responded to the increasing demand driven by the global trend toward carbon neutrality through the development of the RTGMS, which leverages our long-standing expertise in gas sensing technology. The case studies presented in this paper represent only a portion of the system's potential applications. With our diverse sensor lineup, RTGMS can deliver customized solutions tailored to on-site requirements. In addition to monitoring gas composition, calorific value, and specific gravity, the system enables real-time measurement of compression factors, methane number, combustion rate, and other critical parameters.

We will continue to advance and promote this integrated monitoring platform—combining the continuous measurement capabilities of detectors with the high accuracy of analyzers—to support a broad range of industrial markets and contribute to the transition toward carbon-neutral operations.

 t pp s/dI o.gpr5dl0gl0(1de: jconfd606) 3(60100(

Interlaboratory Comparison to determine the majority composition of Biomethane

Andreia Fioravante [1*], *Isabela Alves* [1], *Vanderléa de Souza* [1], *Viviane Mello* [1], *Claudia Ribeiro* [1], *Maurício Guimarães*[1], and *Cristiane Augusto* [1]

[1]Instituto Nacional de Metrologia, Qualidade e Tecnologia (Inmetro), Av. N. Sra. Graças 50 Xerém, Duque de Caxias, RJ, Brazil

Abstract. With the energy transition, many studies have been carried out on biogas, as it is considered a renewable energy source. For its use as fuel, a purification process is necessary, becoming biomethane, with a higher concentration of methane and calorific value. The quality control of biomethane from landfills intended different uses is established by ANP Resolution No. 886 of 2022. To comply with this Resolution, it is necessary to correctly measure the composition of the components using high-precision analytical methods. Therefore, the analytical methods were adapted so that they can determine the composition of the major components present in biomethane. A collaborative interlaboratory comparison study was conducted in order to improve and compare measurements composition results for mixtures of biomethane with major components (methane around 90%mol/mol and carbon dioxide around 1.75%mol/mol) with the purpose to assess technical competence of the participants. The study aimed to provide evidence for validated methods using alternative analytical techniques for the proposed measurement using traceable standards, and verify whether the results are in accordance with the parameters reported in the resolution. Through the results, it was possible to conclude that the methods are suitable for their intended use for methane, but for carbon dioxide, there were divergences from the reference value when it was used the portable sensor analysers.

* Corresponding author: aalima@inmetro.gov.br

t p s/dl o.gr5dl0g0(1d: jconf606) 3(60100(

1 Introduction

The increase in energy consumption, together with the limitation of fossil resources, brings to light the need to reduce emissions of pollutants into the atmosphere. In this sense, the debate on the preservation of natural resources, the development of stricter environmental legislation, and even the establishment of global treatments, were responsible for the growth of sustainable appeal. Currently, the use of renewable sources and the diversification of the national energy matrix have become the subject of debates and the focus of research [1],[2]-[3].

In recent years, many studies on the use of renewable energy sources for power generation are being carried out. Biogas is considered a renewable energy source and has gained interest mainly because it is considered low pollutant [4].

The product generated from raw biogas consists of methane, carbon dioxide and other gases such as oxygen, nitrogen, hydrocarbons and siloxanes. It can be produced from agricultural waste, urban solid waste contained in sanitary landfills or from sewage in sewage treatment plants, being, in all cases, under the action of bacteria by anaerobic digestion. After stages of purification of biogas, biomethane is the result, a renewable energy source with a high methane content in its composition [5].

The production and use of biomethane in the world has been growing over the years. It is important to emphasize that several measures to encourage the production of biomethane were launched in Brazil [6].

1.1 Biogas and biomethane scenario in Brazil

The Brazilian biogas market for energy applications has expanded significantly. Brazil has 1,365 biogas production plants, and between 2022 and 2023, 338 new units were implemented, representing a 32% growth in the total number of plants and a 21% increase in biogas volume, reaching 4.15 billion Nm^3/year. Regarding its application, 86% is destined for electricity generation, 3% for biomethane, and 9% for thermal energy. The plants that started operating in 2024 have an average installed capacity 54% greater than those that started operating in 2023 [7].

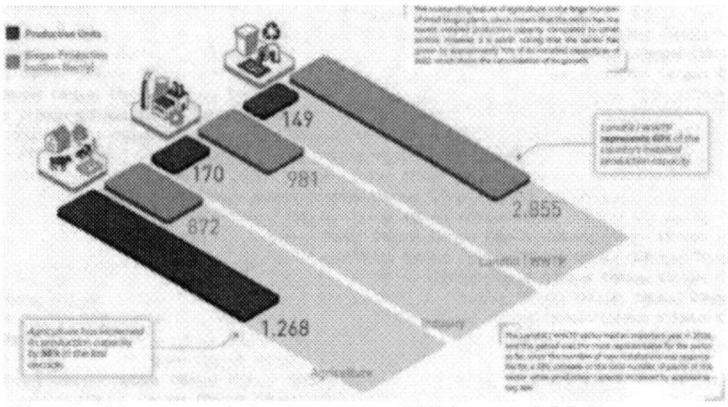

Fig. 1. Biogas origin (source: CIBiogas, 2024)

Fig. 2. Biogas relation between origin and demand (source: CIBiogas, 2024)

In order to guarantee the quality of biomethane, the Brazilian Oil and Gas Agency (Agência Nacional de Petróleo, Gás Natural e Biocombustíveis -ANP) through Resolution No. 886 of September 29, 2022, established the criteria for approval of quality control and the specification of this energy source from landfills and sewage treatment stations for vehicular use and residential, industrial and commercial installations to be sold throughout the Brazilian territory [8].

However, complying with industry regulatory specifications can prove challenging for laboratories, as there is no specific analytical method to determine the composition of biomethane, suggesting the use of methods described in standards for natural gas analysis, such as: NBR 14903, ASTM D 1945, and ISO 6974 [9]-[11]. Nevertheless, the composition of these gases is not similar to biomethane gas, nor are the concentration limits. Despite this, the analytical methodologies described in the current resolutions, similar to those established for natural gas, with additional parameters such as halogenated compounds and siloxanes, can represent technical and economic challenges for medium and small-scale producers, due to the need for high-cost equipment and specialized labor. Furthermore, there is a limitation in the number of national laboratories capable of performing these analyses in Brazil. Currently, there are no laboratories accredited for biomethane testing; there is only one accredited laboratory for natural gas, which can partially perform some methodologies.

Therefore, as biomass represents almost one-third of Brazil's domestic energy supply, giving Brazil the cleanest energy matrix, the application of gas metrology is essential to ensure reliable and competitive quality control results for this energy input [12].

1.2 The importance of scientific metrology in biomethane regulatory resolutions

In this context, pilot-scale laboratory applied research is necessary to demonstrate the possibility of adjustments to the analytical parameters of the aforementioned resolutions, in relation to both quality requirements and the use of simpler and more agile alternative methodologies that guarantee the required technical safety. This regulatory improvement is essential to consolidate biomethane as a strategic fuel,

t ps/di o.gr5d0g0(1d: jconf606) 3(60100(

boosting investments and strengthening the low-carbon economy in Brazil.
Inmetro, the Brazilian Metrology Institute (Inmetro), through its Gas Analysis Laboratory (Lanag), develops and validates analytical methodologies in the areas of energy and environmental control, being the main entity responsible for disseminating metrological knowledge of gases in Brazil. It produces certified reference materials, works on the transfer of metrological traceability, and coordinates laboratory comparison programs. It stands out for its highly complex equipment park, accessible to external users, and operates in accordance with the ISO 17025 and ISO 17034 standards [13],[14]. The experience in developing traceability service and products for biogas and biomethane measurements will drive technical and scientific advances, market growth, professional qualification, and the transfer of metrological knowledge, contributing to national development and increasing the reliability of measurements.
The study presented in this paper aims to ensure the reliability of research through rigorous evaluation of performance parameters, complying with current regulatory resolutions and proposing new, more agile and effective analytical methodologies. These methodologies will serve as a basis for proposing regulatory changes.

1.2.1 Biogas Key-comparison

One of the tools for control and guarantee of the validity of analytical results is the use of certified reference materials (CRMs), which are essential for validation of methods, as well as in the calibration of instruments ensuring the metrological traceability, reliability and comparability of results in order to guarantee the quality of measurements carried out. In order to evaluate the confidence of these standards, National Metrology Institutes, including Inmetro, are required to participate in international key-comparisons (KC). The first biogas KC, CCQM-K112, was conducted in 2020, with the following amount of fraction of 7 components gravimetrically prepared (Fig. 3).

Component	Amount fraction x (cmol mol^{-1})
Methane	40 – 56
Carbon dioxide	36 – 42
Nitrogen	12 – 16
Hydrogen	0.8 – 1.2
Oxygen	0.3 – 0.6
Ethane	0.02 – 0.08
Propane	0.005 – 0.020

Fig. 3. Specification of standards

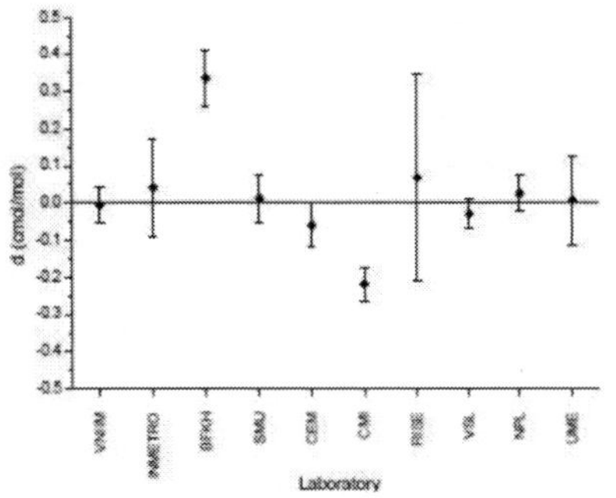

Fig. 4. Degree-of-equivalence for Methane

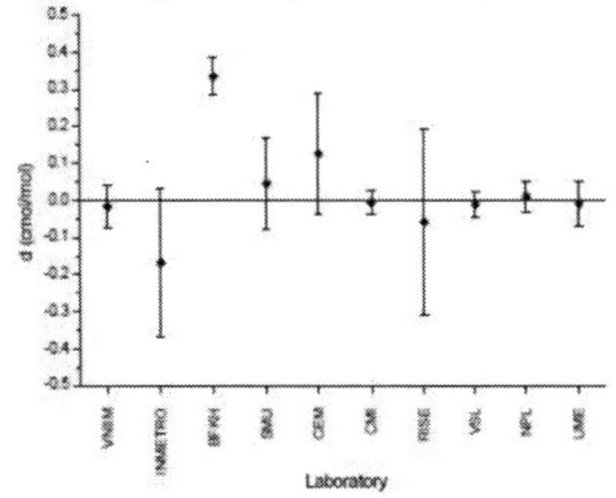

Fig. 5. Degree-of-equivalence for Carbon dioxide

Inmetro's results were satisfactory for all biogas components, according to results presented for some of the main components of biogas at Figure 4 and 5 [15].

1.2.2 Biomethane collaborative interlaboratory comparison study

An Interlaboratory Comparison (IC) aims to compare measurement results obtained under similar conditions from two or more laboratories and to evaluate the technical competence of the participants. Generally, it assesses laboratory performance, highlights problems related to testing procedures, and enables corrective and/or preventive actions to be taken, with the specific objective to contribute to the improvement of measurements focused on the renewable energy sector.
A collaborative interlaboratory comparison study was conducted in order to improve and compare measurements composition results for mixtures of biomethane with major components (methane around 90% mol/mol and carbon dioxide around 1.75% mol/mol) with the purpose to assess technical competence of the participants.
That way, specific methods for determining the composition of biomethane will make it possible to properly verify the composition of the fuel, making it of high quality, that is, high calorific value, generating more energy. The methodology will also contribute to generating benefits for industrial development, the environment, the economy and society.

2 Methods and Development

Ten cylinders containing synthetic mixtures of methane (approximately 90 mol/mol concentration) and carbon dioxide (approximately 1.75 mol/mol concentration)

 t pp s/dl o.gpr5dl0gl0(1de: jconfd606) 3(60100(

were available. Inmetro was responsible for the reference value of the samples and data compilation, through the calibration of each gravimetric standard of biomethane acquired from an accreditated body industry against Inmetro's own PRM. The analythical technique used by Inmetro in calibrating the standards was micro GC-TCD (model Agilent 490), according to ISO 6143:2024 [16]. Inmetro's certified reference material from biomethane were produce according to ISO 6142:2001 [17].

The participants used their own analytical methodology (Table 1) to determine the majority composition of biomethane. This study was conducted in order to provide evidence for the development of national standards using alternative analytical techniques for the proposed measurement.

Table 1. Analythical techniques.

Lab ID	Technique
1	GC-TCD
2	GC-TCD
3	Micro GC-TCD
4	Micro GC-TCD
5	GC-TCD
6	NDIR
7	NDIR
8	Portable analyser - Electrochemical sensor
9	Portable analyser - Electrochemical sensor

The tolerance limit for methane and carbon dioxide was established following National ISO Standard for natural gas [9], which states that for components with concentrations of 1 to 10 %mol/mol, the limit is 10%, and for components with concentrations of 50 to 100 %mol/mol, it is 3%. We have decided to use half of the tolerance limit from the Standard.

3 Results and discussion

The results of the comparative study of analytical methodologies are summarized in Tables 2 and 3, followed by the graphical results showing the degrees of equivalence (Figures 6 and 7) [18].

Table 2. Methane results.

	Inmetro/ Methane (%mol/mol)	Lab/ Methane (%mol/mol)	Absolute deviation
Lab 1	90,59	90,97	0,38
Lab 2	90,65	90,46	-0,19
Lab 3	90,65	90,97	0,32
Lab 4	90,71	91,98	1,27
Lab 5	90,55	90,79	0,24
Lab 6	90,54	90,53	-0,01
Lab 7	90,66	91,10	0,44
Lab 8	90,62	90,86	0,24
Lab 9	90,07	89,91	-0,16

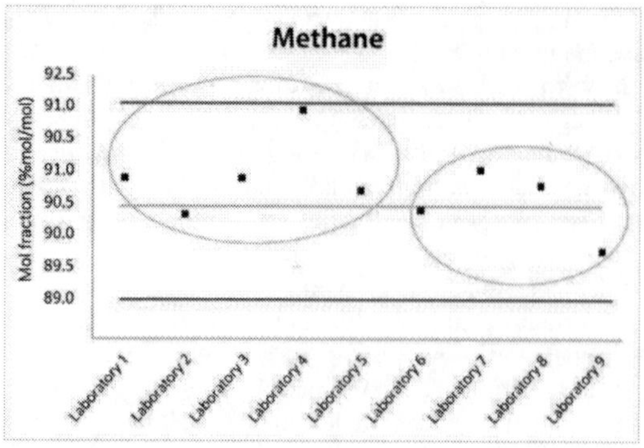

Fig. 6. Degree-of-equivalence for Methane

The results of the comparison of methodologies for the component methane, whose concentration is around 90 %mol/mol, proved satisfactory. All laboratories (Fig. 6) obtained satisfactory performance, both using the gas chromatography analytical technique and the portable and fixed analyzer techniques.

Table 3. Carbon dioxide results.

	Inmetro/ CO_2 (% mol/mol)	Lab/ CO_2 (%mol/mol)	Absolute deviation
Lab1	1,75	1,76	0,01
Lab 2	1,76	1,74	-0,02
Lab 3	1,75	1,68	-0,07
Lab 4	1,75	1,73	-0,02
Lab 5	1,76	1,75	-0,01
Lab 6	1,75	1,80	0,05
Lab 7	1,74	1,70	-0,04
Lab 8	1,75	1,56	-0,19
Lab 9	1,71	1,91	0,20

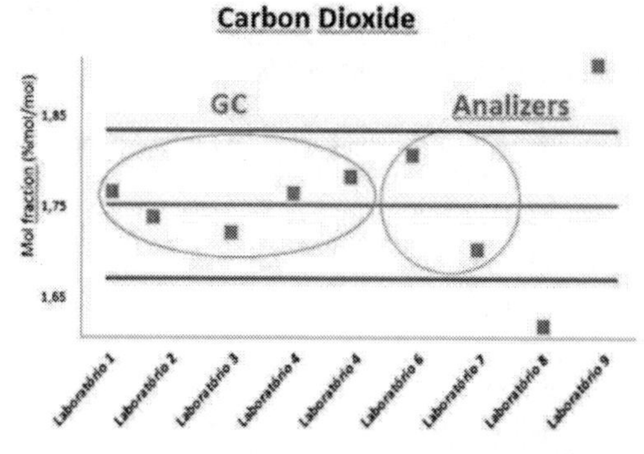

Fig. 7. Degree-of-equivalence for Carbon dioxide

The results of the comparison of methodologies for the component Carbon dioxide, whose concentration is around 1.75 %mol/mol, showed satisfactory performance for all 05 laboratories that used the gas chromatography analysis technique. For the laboratories that used the fixed analyzer technique, 02 laboratories had satisfactory performance, while the 02 laboratories that used the portable analyzer technique had unsatisfactory performance.

The reason for this could be evaluated as:

- Absence of a certified reference material (CRM) close to the sample measurement range during instrument calibration;
- Use of expired CRMs or CRMs from manufacturers without recognized competence;
- Lack of experience in performing auto zero gas monitoring.

4 Conclusion

The use of biomethane aims to promote an energy transition aligned with Brazil's potential and competitive advantages, consolidating the country as a protagonist in the national and international energy scenario.

All results for methane were satisfactory. However, the participants that measures carbon dioxide using portable analyzers had unsatisfactory results, which can be explained by misuse or even lack of certified reference materials in the calibration of the sensors.

A proficiency test is being organized in order to evaluate a major number of laboratories including the estimative of the uncertainty of the measurements with the use of Inmetro´s primary reference gas mixtures of biomethane.

Acknowledgment

The authors would like to acknowledge to the Fundação de apoio a UFMG (Fundep) for the support related to the project "Ensaios químicos".

References

1. International energy agency, Outlook for biogas and biomethane - a global geospatial assessment, IEA, March 2025. Online [Accessed January 16, 2026]
2. Programa de Energia para o Brasil (PEB), Biogás no Brasil - Potencial de descarbonização a curto prazo, Technical report 01-2022. São Paulo: Instituto 17, 2022. Online [Accessed January 16, 2026] [In Portuguese].
3. World BiogasAssociation-WBA, GlobalPotential of Biogas, 2019.
4. Fernandes, G.; Mariani, L. O alto potencial de produção e uso fará do biogás a próxima fronteira da energia renovável no Brasil? FGV Energia, 2019.
5. Van der Veen, A. M. H.; Brown, A. S.; Li, J.; Murugan, A.; Heinonen, M.; Haloua, F.; Arrhenius, K. Measurement requirements for biogas specifications. EDP Sciences, 2015. DOI: https://doi.org/10.1051/metrology/20150008006
6. Government of Brazil. Incentive to the production and sustainable use of biomethane, 2022. [Accessed August 5, 2022] [In Portuguese].
7. Centro Internacional de Energias Renováveis – Biogás Panorama do Biogás no Brasil 2024. CIBiogás (Brasil) Relatório Técnico nº 001/2024 – Foz do Iguaçu, CIBiogás, 2024. Online [Accessed January 16, 2026] [In Portuguese].
8. National Agency of Petroleum, Natural Gas and Biofuels (ANP), Biomethane, 2017. [Accessed April 7, 2021] [In Portuguese].
9. ABNT NBR 14903:2014 - Gás natural – Determinação da composição química por cromatografia em fase gasosa.
10. ASTM D1945-14:2019 - Standard Test Method for Analysis of Natural Gas by Gas chromatography.
11. ISO 6974-1:2012 - Natural gas — Determination of composition and associated uncertainty by gas chromatography - Part 1: General guidelines and calculation of composition.
12. Energia e Biogás. Panorama of electric power generation from biogas sources, 2021. [Accessed August 27, 2022] [In Portuguese].
13. ABNT NBR ISO/IEC 17025:2017 – Requisitos Gerais para a competência de laboratórios de ensaio e calibração.
14. ABNT NBR ISO/IEC 17034:2017 - Requisitos gerais para a competência de produtores de material de referência.
15. Adriaan M H van der Veen *et al.* International comparison CCQM-K112 biogas. Metrologia, Volume 57, Number 1A, 2020. DOI: 10.1088/0026-1394/57/1A/08011
16. ISO 6143:2025 - Gas analysis — Comparison methods for determining and checking the composition of calibration gas mixtures. Ed.3.
17. ISO 6142-1:2015 - Gas analysis — Preparation of calibration gas mixtures, Part 1: Gravimetric method for Class I mixtures, Ed.1.
18. Inmetro. Technical report R-0786/2024 - Estudo Colaborativo de Comparação de Metodologias Analíticas de Determinação de Composição de Biometano, 2024.

https://doi.org/10.1051/epjconf/606(3560100(

Quantifying fugitive hydrogen emissions at the component scale: High-flow sampling method

Haydn Barros [1*], *Hannah* Cheales-Norman[1], *Aimee* Buck[1], *Jon* Helmore[1] and *Nigel* Yarrow[1].

[1]National Physical Laboratory, Atmospheric Environmental Science Department, Hampton Road, Teddington, TW11 0LW, UK

Abstract. Hydrogen is a key energy vector in the transition to net zero with its industrial use expected to grow significantly in the next decades. At present there are no standard methods specific for hydrogen fugitive emissions quantification, and these are required to inform monitoring schemes and support regulations. On the other hand, the high flow sampling method (HFS) has been successfully used for natural gas leak rate quantification in the industry. So, in this work we present for the first time a performance evaluation of the HFS method for quantifying hydrogen leak rates.

1 Introduction

The momentum behind hydrogen is strong. In total, 60 governments (including the European Union) have adopted H_2 strategies [1] The current H_2 global market is valued at around $283 billion in 2025, while projections estimate the market could reach $556 billion by 2034 [2]. Experience in the natural gas industry has found fugitive emissions in different scenarios/scales along the whole production chain. The H_2 leaks could be even more significant, since it is a lighter molecule, is less viscous, and some H_2 storage systems work at higher pressures as compared to natural gas facilities. For those reasons, great efforts are being devoted to better understand the potential H_2 emissions and to identify and develop the reliable methods for monitoring those. There are several international initiatives aiming to fill the existing gaps in the H_2 emissions monitoring [3,4,5,6]. A list of their main topics follows.

- Hydrogen emissions research needs
- Detect and quantify hydrogen emission
- Identify / develop tools and technologies (BAT)
- **Develop methods**, protocols, guidance
- Harmonize existing practices across the value chain
- Accurate hydrogen emissions inventories
- Fill up gaps where little empirical data exist
- Move towards standardisation
- Support regulators and industry
- Better impact assessment (environ., risk, economic)

NPL is actively working on the development of practical methods to reliably quantify industrial hydrogen fugitive emissions [7]. In the present work we test the High-Flow Sampling method (HFS) and assess its feasibility to monitor industrial H_2 leaks at component scale (individual components). The main challenges to quantify H_2 leaks and the found solutions are presented.

2 The High-Flow Method

The method consists in sampling at a high flow rate (above hundred litres per minute) using a hover type instrument capable of drawing in a considerable amount of ambient air along with the escaping gas from the leaky component. Then the concentration of the diluted gas is measured, and the leak rate is calculated. The method has been used for many years in the industry to monitor natural gas leaks, so commercial devices incorporates natural gas (CH_4) sensors internally.

When using a portable (back pack) High-Flow Sampler, the inlet tube is approached to the leak source (~ 1 cm) sucking the leaked gas inside the sampler. The leaked gas will be diluted with ambient air (the sampling flow rate (S) is larger than the leak rate). Then, the sampling flow rate and the gas volumetric concentration (c) inside the instrument are measured, and the leak rate (volumetric rate, Q_V) is calculated as follows:

$$Q_V = S\,(c - BG) \qquad (1)$$

BG represents the background gas concentration corresponding to a far enough point in which the leak source have no effect. Here Q_V and S are expressed in L/min while c and BG are volumetric fractions. Eq. 1 assumes that the gas is pure (c = 1 or 100%), so the lower concentration measured is caused only by the dilution of the gas with ambient air. In the case of diluted or blended gases Eq. 1 must be corrected accordingly. Eq. 1 also assumes that the whole leaked gas is being sampled during the measurement (in steady state). Great care must be taken to ensure the full capture of the leak, otherwise the actual leak rate could be greatly underestimated.

* Corresponding author: haydn.barros@npl.co.uk

https://doi.org/10.1051/epjconf/606(3560100(

In this work we have selected the Hi-Flow Bacharach sampler (BHFS). The concentration of the diluted H_2 inside the HFS was measured by a GS-700-H2 (Gasurveyor - Teledyne) detector. In the lab it was determined that this detector can reliably measure H_2 down to 50 ppm (μmol/mol), which combined with the BHFS capacity to sample 200 L/min allow us to measure hydrogen leaks of 0.01 L/min (if full capture is achieved). On the other hand, although the BHFS is designed to be intrinsically safe for use in hazardous locations (Class I, Division 1, Groups A, B, C & D in North America) is not ATEX rated. For that reason, we tested the instrument for a maximum internal (diluted) concentration below 4% v/v H_2 (100 %LEL), allowing to quantify leaks up to 7-8 L/min.

Fig. 1. High-Flow sampler measuring a leak rate.

Several leak rate tests were performed simulating leak H_2 rates between 0.5 and 7 L/min. When the full capture is guaranteed the method provide high accuracy (deviations < 1%) in all the cases (not shown). It is worth noting that the full capture of the leak is not always possible in real conditions, and this is the most relevant issue in the application of this method.

3 High-Flow Sampling rate

The sampling rate of a High-Flow instrument is generally measured with an orifice plate device. The BHFS's orifice plate was calibrated. The main components of the calibration system are: gas supply (air), mass flow controller (MFC), the certified flow meter (Mesa Labs DryCal 1020 flow meter) and the High-Flow Sampler to be calibrated in terms of flow rate. The BHFS flow meter has good linearity (Fig. 2). However, the three lower flow rates are systematically below the actual flow rate (4-7%) and their uncertainties (2σ) cannot explain this difference. For that reason these were excluded for the linear regression fitting. All other seven flow rates overlaps with the actual value within the 2σ interval. The obtained linear fitting is almost the ideal trend (y = x) and no recalibration was required.

The measured sampling rate values are corrected for the effect of temperature on the flow rate (using the ratio of absolute temperatures). It is worth noting that the BHFS maximum flow rate is ~220 L/min, and the minimum accurate value is 150 L/min (as shows Fig. 2). Generally the BHFS sampling rate (produced by an internal fan) is quite stable (σ < 2% in several minutes).

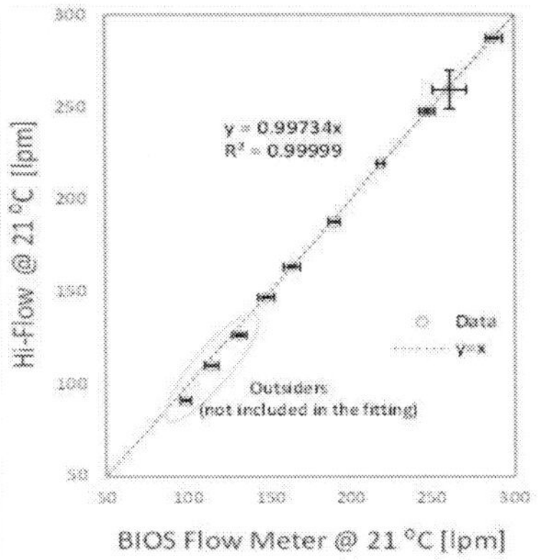

Fig. 2. High-Flow sampling rate calibration.

4 Effect of the hydrogen concentration on the sampling rate measurement

The measurement of the sampling flow rate is an essential value to quantify leaks using the High-Flow method. This is usually measured by an orifice and two sensors providing a differential pressure measurement. Generally, this kind of measuring systems are calibrated with dry air, and the read flow rate is depending on the gas passing through the pipe having the orifice and exerting pressure on the sensors. The equation relating the mass flow rate with the differential pressure also considers the gas type, and if we assume that the flow is below the ultrasonic regime (i.e. the differential pressure is not too high), it can be expressed as follows.

$$S_M = \frac{1}{\sqrt{1-\beta^4}} \frac{\pi}{4} d^2 \sqrt{2\,\Delta p\,\rho} \quad (2)$$

S_M is the mass flow rate, Δp is the differential pressure (before - after the orifice), β is the ratio between the orifice diameter (d) and the pipe diameters (D), and ρ is the gas density. This expression can be converted in the volumetric flow rate just dividing by the gas density

$$S_V = \frac{1}{\sqrt{1-\beta^4}} \frac{\pi}{4} d^2 \sqrt{2\,\frac{\Delta p}{\rho}} \quad (3)$$

However, it is necessary to introduce a discharge coefficient C accounting for geometrical effects, it determines the relation between the real flow through a device and the theoretical flow. With a fixed geometry, C only depends on the Reynolds number of the gas and C can be seen as a calibration constant for the device. As a first approximation we will consider that the behaviour of air and H_2 near normal temperature and pressure (NTP, i.e. 20C and 101.325 kPa) are close enough to consider C to be the same for both gases. This is based on the observation that Reynolds numbers for gas flowing through pipes are usually ~10^4 to 10^7, then the effect on the parameter C is low. In fact, there are parametrizations of C showing that when the Reynolds number tends to infinity then the so-called C_∞ depends only on the geometry [8]. This approximation could introduce discrepancies ≤ 0.01% when changing from air to H_2 (at NTP). In second place, we use an expansion coefficient ε describing the behaviour of the compressible media (gas). Using the same approximation (Reynolds number tends to infinity) the Soltz equation can be used to approximate the expansion coefficient, and it can be shown that it depends on geometrical factors, the differential pressure, the pressure before the orifice and the isentropic coefficient κ *(Cp/Cv)* for the gas flowing through the orifice [8]. At NPT most diatomic gases has a κ value of approximately 1.4, so also the expansion coefficient can be considered the same constant in the case of comparing air (mainly N_2 and O_2) and hydrogen (H_2) flows. This can be simplified using a single factor (F) for all the constant parameters (for a given device).

$$S_V = \frac{C}{\sqrt{1-\beta^4}}\, \varepsilon\, \frac{\pi}{4}\, d^2 \sqrt{2\,\frac{\Delta p}{\rho}} = F\sqrt{\frac{\Delta p}{\rho}} \tag{4}$$

Different gases density will affect the measurement. For a given flow measurement (reading), there is a given Δp. This differential pressure will take place during the calibration (dry air) and when the gas is measured.

$$S_{V\text{-}air} = F\sqrt{\frac{\Delta p}{\rho_{air}}} \quad \text{or} \quad S_{V\text{-}gas} = F\sqrt{\frac{\Delta p}{\rho_{gas}}} \tag{5}$$

Combining both equations of equal Δp, we obtain

$$S_{V\text{-}gas} = S_{V\text{-}air}\sqrt{\frac{\rho_{air}}{\rho_{gas}}} \tag{6}$$

The gas being measured is a combination between the leaked gas (in this case pure H_2) and the ambient air (which we will suppose dry air), then the density of the flowing gas can be expressed as follows

$$\rho_{gas} = \rho_{H_2}\, c + \rho_{air}(1-c) \tag{7}$$

Where c is the measured concentration of the leaked gas (in this case H_2), as in Eq. 1. Any correct reading (measured sampling rate) must correspond to the calibration gas, which is generally air ($S_{V\text{-}read} = S_{V\text{-}air}$). Then, combining Eq. 6 and 7, we obtain a corrected sampling flow rate measurement in terms of the direct read flow rate and the gas concentration measured internally where the orifice is located.

$$S_{V.gas} = \frac{S_{V.read}}{\sqrt{1 + c\left(\frac{\rho_{H_2}}{\rho_{air}} - 1\right)}} \tag{8}$$

The quantification of the hydrogen volumetric leak rate can be done combining Eq. 1 and 8, the result here is simplified for cases with negligible background.

$$Q_{V.H_2} = \frac{S_{V.read}\; c}{\sqrt{1 + c\left(\frac{\rho_{H_2}}{\rho_{air}} - 1\right)}} \tag{9}$$

Any gas density is changing with the temperature, but fortunately the ratio of their densities is almost constant (in this case ρ_{H_2}/ρ_{air}= 14.56). In most cases, the large dilution caused by the High Flow sampling will produce internal concentration values relatively low and the correction factor in the denominator of Eq. 9 could be neglected. However, this could not be the case for large leaks of pure hydrogen, and it is worth having a quantitative estimation of this correction. Fig. 3 shows the magnitude of the correction factor (k) to be applied as a function of the hydrogen concentration measured in the internal pipe of the system where the measuring orifice is located.

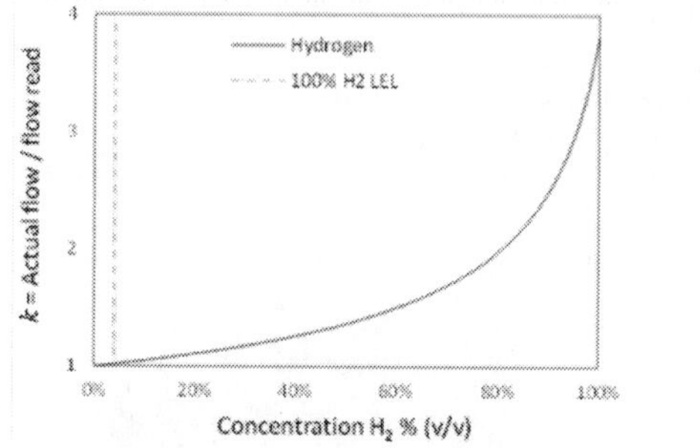

Fig. 3. Correction for the leak rate measurement as a function of the measured internal H_2 concentration (c).

In the case of common leaks, the concentration inside the sampler is usually below the lower explosivity limit (LEL). Fig.4 shows the details within this range.

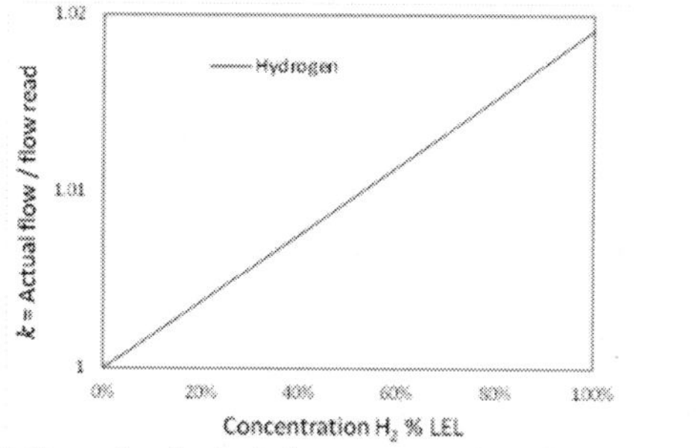

Fig. 4. Correction for the leak rate measurement for moderate H_2 leaks ($c \leq$ 100% H_2 LEL).

For instance, in the case of measuring 50% LEL, the flow rate must be corrected by a factor of $\approx$ 1.01, i.e. the flow of the actual gas is $\approx$ 1% higher than an air flow producing the same reading (the same differential pressure through the same device). Finally, the emitted leak rate can be quantified as

$$Q_{V.H_2} = k\, S_{V.read}\; c \tag{10}$$

Where k is a function of the measured hydrogen concentration (as shown in Fig. 3). It is often useful to represent the leak rate in terms of mass flow rate (Q_M) instead volumetric flow rate, this is important for emissions inventories and other cases where total emissions or balances are relevant. This can be calculated using the ideal gas state equation.

$$Q_M = Q_V\, \frac{M\,P}{R\,T} \tag{11}$$

Where M is the molar mass of the target gas (M = 2 g/mol in the case of H_2), P and T are the absolute pressure and temperature where the measurement is done (Pa and K), and R is the ideal gas constant (R $\approx$

 https://doi.org/10.1051/epjconf/606(3560100(

8,314 L Pa/mol K). Then with a Q_V (L/min) a mass leak rate Q_M (g/min) can be calculated.

5 High-Flow sampling method for H_2 tested in outdoor windy conditions

Sampling efficiency depends on several parameters associated with the complex aerodynamic features of the leak jet and its capture process due to the drag air stream sucked in by the HFS. There are situations in which turbulence is created at the entrance of the sampling hose, dispersing the H_2 gas randomly.

The leak profile has a region where the gas linear momentum dominates the transport, a transition zone and a diffusional zone. The fast zone features is given by the leak rate, the leak source size and geometry). Depending on the leaked gas speed at the interface, the jet could be difficult to capture (unless the sampling hose is perfectly aligned). It is important to remind that in a real situation it is necessary to detect the leak and determine the source location as accurately as possible, to then approach the leak source without known the jet direction and overcoming any physical restriction (pipework complexity and geometry).

We have simulated hydrogen leaks using open end pipes with different apertures (0.1; 0.3; 0.5; 1 and 4 mm) and leak rates from 0.5 to 5 L/min in outdoor windy conditions. One relevant parameter is the relative orientation of the sampling hose with respect to the leak jet and or the wind direction. This last was tested. Fig. 5 shows one of these configurations.

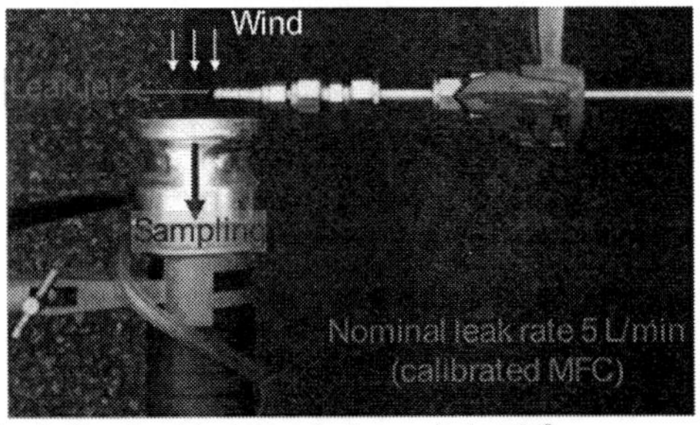

Fig. 5. Simulated hydrogen leak sampled at 90°.

During the measurement the sampling rate was stable, however, in some rare cases, the measured H_2 concentration was variable, due to turbulent flow in the entrance of the hose. Fig. 6 shows the case exhibiting more variability in the measured concentration. In this example the measurement does not stabilise but fluctuates, and only the maximum concentration values produce the right (nominal) leak rate calculation. For that reason, the measurement needs to be verified for stability, it shall be a steady state process. If this is not possible, then it might be advisable to choose only the maximum values (spiking) to avoid underestimating the actual leakage rate. The physical explanation is that only in brief occasions the leaked gas is fully captured by the sample in the conditions of this specific test.

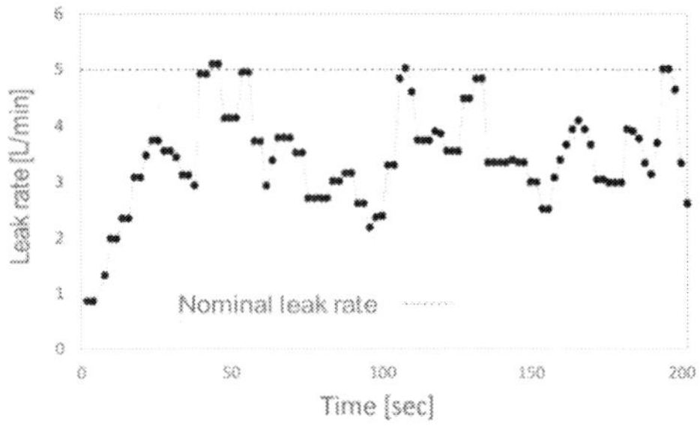

Fig. 6. Measured leak rate in extreme windy conditions.

In another test series the leak jet was facing upwards (vertical) and the sampler was placed in different orientations respect to the wind (Fig. 7). The full capture condition was checked (inserting the pipe end in the sampler hose), several inlet attachments were tested (cone, hose), and the sampling distance was variated (0, 1 and 5 cm from the interface) simulating not knowing the precise location of the leak source.

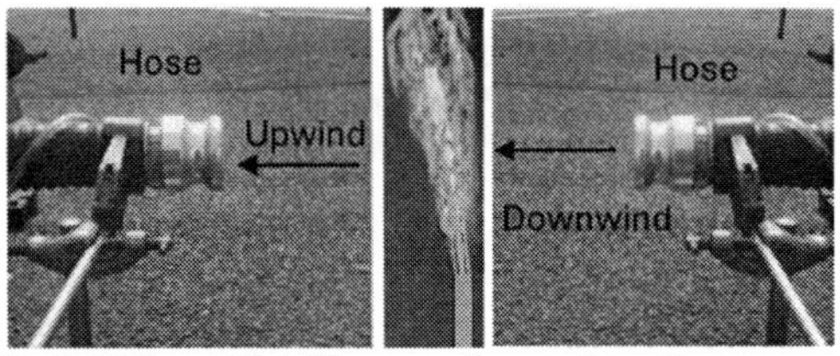

Fig. 7. Different High-Flow sampling configurations.

All the measurement performed at the full capture condition coincides with the nominal simulated leak rate. Also, in several simple cases the measured values are close to the actual leak rates. However, in many cases there are large discrepancies, mainly underestimating the leak rate due to a lack of sampling efficiency. For a sake of brevity, here it is shown only one case, which is depicted in Fig. 8. In that case, the measurement from *Above* (sampling hose aligned with the leak jet) works well in close proximity (0 cm) or at 1 cm of the leaky orifice, but at a distance of 5 cm the values are more dispersed and underestimate the leak rate in half or more. At 0 cm also the *Downwind* sampling provides a good estimate of the leak rate. The other measurement conditions systematically underestimate the actual leak rate and produce values with a larger dispersion.

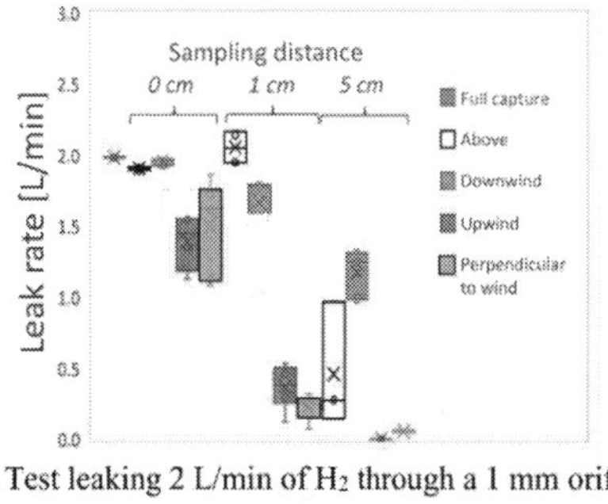

Fig. 8. Test leaking 2 L/min of H_2 through a 1 mm orifice.

https://doi.org/10.1051/epjconf/606(3560100(

In real cases it is not possible to know the sampling efficiency beforehand, therefore it is recommend performing two measurements, one at the highest sampling rate and another one at a lower rate (60-80% of the highest sampling rate). If in both cases the obtained results are close enough (< 5 or 10%) they can be regarded as efficient samplings, otherwise it will imply a lack of gas collection. In those cases, it is advisable to use a different sampling strategy, as using the High-Flow sampling through a bag type enclosure (antistatic) to trap the escaping gas and repeat the measurements at two different rates. In the case of flanges, it is better to use a plastic strap (also antistatic) with inlet and outlet apertures to drag the leaking gas.

6 High-Flow Sampling using bags or other encapsulations

Generally, when measuring with a HFS, the readings grow continuously (transient sampling) until a plateau is reached, in this steady state (ideal) condition all the gas being leaked is trapped by the HFS at a constant sampling rate, and the valid measurements shall be taken in this situation. In the case of using bags or other kind of encapsulation to constrain the leaked gas plume, for a short period of time the encapsulation could built up a certain amount of gas, causing a peak before the plateau. Fig. 11 gives an example of these situations.

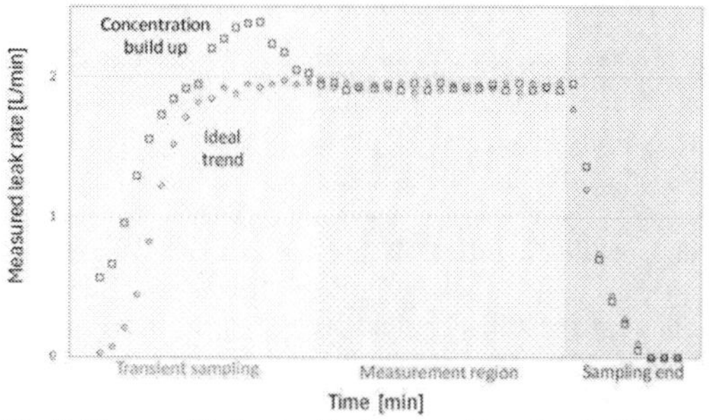

Fig. 9. Measured leak rate, ideal case and transient build up.

In these cases, it is of paramount importance waiting until the readings reach a plateau to avoid leak rate overestimations. Measurements shall be taken only from the plateau (measurement region), discarding the readings from the transient and end sampling regions. The values in the plateau should correspond to the actual leak rate, unless a sub-optimal sampling condition resulted from limitations in the positioning of the sampling tip (hose, cone, etc.), as shown in Fig. 8.

For a sampling rate ~200 L/min it is advised to wait at least 1 minute until the readings stabilises to an approximate steady state value (plateau), which should be within ±10%. The measurement region should take some 1-2 minutes. In general, at least three readings shall be taken reporting the average value and an expanded uncertainty with a coverage factor k=2 (2 x standard deviations).

7 Integration of HFS and H_2 gas detectors for leak rate quantification

Nowadays there are no available HFS for the quantification of hydrogen fugitive emissions. Fortunately, the two selected instruments (BHFS and GS-700-H2) can record and store data automatically during the measurement. The operator has then two options, use the read values in the devices display, or record the two set of data separately. On the other hand, it is clear that the method (Eq. 1) relies on measuring sampling flow rates and diluted H_2 concentrations simultaneously. In order to evaluate the impact of the data acquisition method we have recorded the data manually and automatically, to be able to compare both. Figure 10 shows the relative differences between the automatic and the manually recorded sampling rates for a set of measurements.

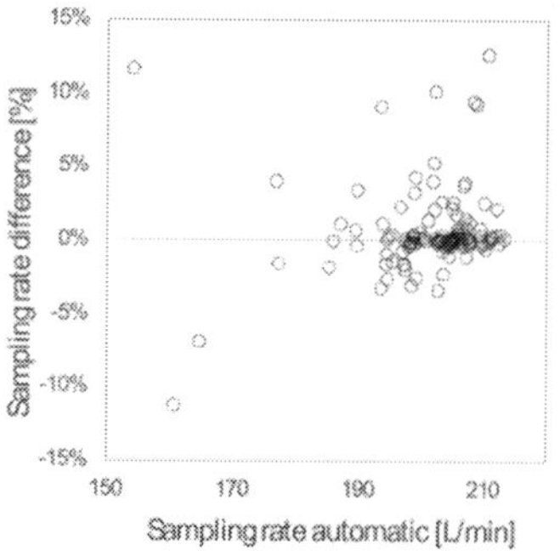

Fig. 10. Relative difference between manual and automatic High-Flos sampling rate measurements.

The differences observed in Fig. 5 are generally below ±5%, with some outliers reaching up to ±12%. Fig. 11 compares the H_2 concentration measurements recorded automatically vs the values recorded manually from the display.

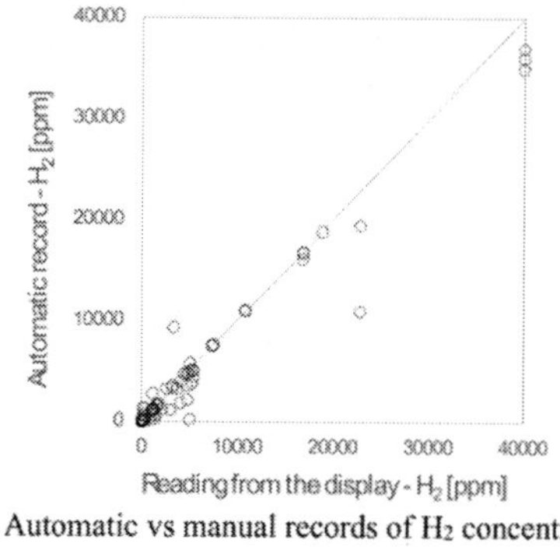

Fig. 11. Automatic vs manual records of H_2 concentrations.

Again, in most cases both automatic and manual recorded values are in close agreement, however, there differences become more important for concentrations below ~5.000 ppm, and even at higher concentration it can be found several outliers. This emphasise the need to automatically register the data and process it statistically to enhance its reliability.

When both measurements are combined, the resultant uncertainty could increase due to a sub-optimal data processing. This evidence the need to integrate both instruments, a HFS and a H_2 gas detector to allow accurate hydrogen leak rate quantifications.

The High Flow Sampling method can be used during walkover surveys to quantify industrial hydrogen fugitive emissions, gathering data on the magnitude of these emissions at component scale, identify common leaky components and their leak distribution, identify concerning leaks and inform LDAR schemes. Also, it is a useful tool to contribute to calculate estimated emission levels (e.g. considering the non scanned components) and could be used as a tool for regulatory compliance.

8 Conclusions

Emission rates can be accurately quantified by high flow sampling if full capture of the hydrogen leaks is achieved. Sufficiently good estimations of the actual hydrogen leak rates can be done in most cases, but it requires to be verified by measuring at two different sampling rates.

The used High Flow system is not ATEX rated for H_2, then the max. measurable H_2 leak rate is ≈ 7-8 L/min. With the used instruments, the lower measurable H_2 leak rate is ≈ 0.01 L/min.

Challenging conditions, as windy scenarios or complex pipe geometries preventing a close approximation to the leak source (even at the 1 cm scale) may cause measurements underestimations

Selecting the appropriate sampling accessory (hose, cone, etc.) is relevant, while in some cases sampling requires full encapsulation (capture bag or flange straps)

In case of variable sampling (due to wind or gas build up) a method refinement in the data analysis and data quality control are both needed: Select only the maximum values in case of large variations caused by wind and select plateau region in case of buildup caused by encapsulation.

Sampling rates and H_2 concentrations records shows differences in the manual and automatic records. Full integration of the two devices is required.

Produced quantitative data is relevant for identifying common leaky components and leak distributions, inform LDAR schemes, estimate emission levels originated from leaky components and as a tool for regulatory compliance.

Acknowledgements: This work was partially funded by the UK Department for Energy Security and Net Zero and the NHyRA project (co-funded by the European Union).

Authors contributions. H. Barros: Theoretical development, experimental design, testing leader, data analysis and charts, work conception and writing. H. Cheales-Norman and A. Buck: logistics, calibrations, testing, data acquisition. J. Helmore and N. Yarrow NPL leaders of projects that partially funded the work.

References

1. IEA, 2025. https://www.iea.org/energy-system/low-emission-fuels/hydrogen.
2. PE&PE, 2025. Precedence Energy and Power Experts. Hydrogen Market Driving the Clean Energy Transition with Innovation and Global Expansion. Report Code 5195. https://www.precedenceresearch.com/hydrogen-market.
3. NHyRA Project. pre-Normative Hydrogen Releases Assessment https://nhyra.eu/
4. UK-HyRES. UK Hub for Research Challenges in Hydrogen & Alternative Liquid Fuels. https://ukhyres.ac.uk/
5. Decarb Project. Metrology for decarbonising the gas grid https://www.decarbgrid.eu/
6. Environmental Defense Fund. Measuring hydrogen emissions: A global research collaboration. https://www.edf.org/measuring-hydrogen-emissions-global-research-collaboration
7. A. Connor*, H. Barros, R. Robinson, A. Guzzini, M. Pellegrini, C. Saccani, J. Holewa-Rataj, T. Kuchta, H. Wigger, T. Vogt, V. Bescos Roy, V. Krohl. Methods for detecting and quantifying hydrogen emissions over a wide range of temporal and spatial scales: a state-of-the-art review. Measurement: Energy. **8**. 100069, 19 p (2025). https://doi.org/10.1016/j.meaene.2025.100069.
8. Y. Pistun. L. Lesovoy. Calculation of expansibility factor of gas at its flow through an orifice plate with flange pressure tappings. JEECS. **2**(2), 80 p. (2016) https://doi.org/10.23939/jeecs2016.02.033A

 https://doi.org/(0.(05(/epjconf/10163510100(

New applications and insights using an old technique

Stefan Persijn *

VSL, Thijsseweg 11, 2629 JA Delft, The Netherlands

Abstract. It has been almost 60 years since the first commercial high-resolution FTIR spectrometer was launched. In gas metrology, such FTIR spectrometers have traditionally been used for a few niche applications, but they have never become a real workhorse. The interest in FTIR has recently revived thanks to new measurement challenges involving multiple reactive gases in applications like CCUS, biogas, and hydrogen quality analysis. However, standard commercial FTIR equipment, such as the gas cell, is typically not fit for purpose for these applications. This paper will discuss some of the necessary modifications to gas cells to exploit the full potential of FTIR as a versatile tool for the selective measurement of reactive gases. Further, common pitfalls in spectral data analysis are discussed. Experimental results on reactive gases in NO_2 and CCUS gas standards are presented to show what can be learnt from FTIR measurements. The paper will conclude with an outlook on whether there is a future for FTIR spectrometers with the recent advent of broadband laser spectrometers with similar multicomponent measurement capabilities.

1 Introduction

In 1842, the Swiss physicist Jean-Daniel Colladon invented the 'light fountain', where light is trapped in a water jet due to total internal reflection of the light at the water-air surface. This invention was waiting a long time for a good application. Nowadays, optical fibres, based on the same principle of total internal reflection, are used to transport huge amounts of data as part of the Internet.

A somewhat similar situation occurred for the use of FTIR spectrometers in gas analysis. In the 1960s the first commercial FTIR analysers became available, and many labs active in gas analysis obtained an FTIR but typically only used it occasionally. For instance, some NMIs operate it for purity analysis of the pure gases used to make reference gas standards. However, more intensive use of the FTIR in gas analysis only occurred in the last 15 years with the emergence of new analytical challenges in applications like hydrogen quality (ISO14687) [1], Carbon Capture Utilisation and Storage (CCUS) [2] and biogas [3]. In these applications, multiple, often reactive gases need to be measured, which is 'ideal' for FTIR spectrometers.

This paper discusses the optimisation of the FTIR measurement cell and precautions on the use of spectral databases and, finally, shows what can be learnt from applying FTIR to the analysis of NO_2 gas standards and gas standards for CCUS.

2 Optimising FTIR cells

There are many providers of absorption cells for FTIR gas measurements. These cells can be either single-pass (providing a path length of typically 10 cm) or multi-pass (providing a path length of typically 2.5 m up to 100 m). The single-pass cells are normally equipped with large optical windows (38 mm - 47 mm) so that they can be used with relatively large infrared beam diameters. As a result, commercial single-pass FTIR cells have a large internal surface area and volume, making them less suitable for analysis of reactive gases, in particular in static reference gas standards.

At VSL we developed an optimised single-pass cell (Figure 1) for analysis of reactive gases.

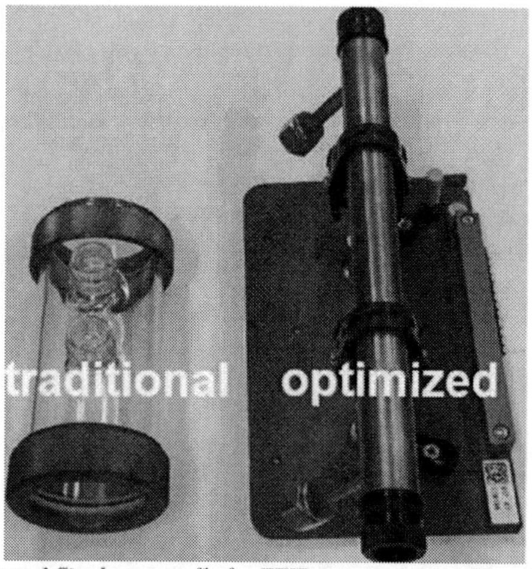

Figure 1 Single-pass cells for FTIR measurements. The cell on the left is a traditional 10 cm cell, with a relatively large volume. The cell on the right has been optimised for the analysis of reactive gases by minimizing the cell volume and use of SilcoNert 2000 passivation.

* Corresponding author: spersijn@vsl.nl

https://doi.org/(0.(05(/epjconf/10163510100(

A comparison of the specifications of both cells is given in Table 1. The optimised cell has a two times longer optical path length, making it about twice as sensitive, yet with a much smaller volume and about twice the lower surface area. This results in a fast stabilisation and less use of reference gas standards. An additional benefit of the optimised cell is that the optical windows are much cheaper due to their small diameter of ½ inch.

Table 1. Comparison of the traditional and optimised cells shown in Figure 1.

	Traditional	Optimised	Traditional / Optimised
Volume (ml)	146	16	9.3
Surface area cell wall (cm^2)	139	63	2.2
Optical path length (cm)	10.5	20.1	0.52
Passivation	No	SilcoNert 2000	

When higher sensitivity is required, multi-pass cells are used. Traditional multi-pass cells are not optimised for analysis of reactive gases or small gas samples. Only in recent years, a few manufacturers have developed optimised cells achieving a path length of 10 m with a modest cell volume of 0.5 L as compared to 2 L for traditional cells (see e.g., [4]).

3 Use of spectral databases

To determine the amount fraction of a certain gas in a sample by FTIR, one can use either a set of reference gas standards or rely on the use of spectral reference data. The latter option is often the only practical approach for reactive gases like nitrous acid (HNO_2) and nitric acid (HNO_3) for which no reference gas standards are available. For the spectral reference data of small molecules, the line strengths from databases such as HITRAN (*hi*gh-resolution *tran*smission molecular absorption database), GEISA (Gestion et Etude des Informations Spectroscopiques Atmosphériques: Management and Study of Atmospheric Spectroscopic Information), or EXOMOL are used. For larger molecules, absorption cross-sections are used, which can be obtained from e.g., the PNNL Quantitative Infrared Database for Gas-Phase Sensing, the NIST Quantitative Infrared Database, and HITRAN.
The most widely used database is HITRAN, which contains millions of line strength data for tens of molecules and every 2 years an update appears. This compilation started more than 50 years ago, and large efforts are made to ensure that the data is reliable [5]. However, these and other spectral reference data are typically not SI-traceable. There are some efforts ongoing to increase the amount of traceable line strength data [6] and a CCQM/GAWG Task Group on Advanced Spectroscopy has recently been established to increase and coordinate this [7].

Care must therefore be taken when using spectral data from the databases. Data from different databases can differ (much) more than is expected based on the reported uncertainties of each database [8]. As an example, Figure 2 shows part of the infrared spectrum of NO_2 from the PNNL and HITRAN databases. There is good agreement for most wavelengths, but around 1263 cm^{-1}, the PNNL database is too high due to the presence of an impurity, in this case HNO_2. Doing such a comparison will show that there are multiple significant differences between the databases (see e.g., [9]). Therefore, users should cross-check the data from different databases before using them and, whenever possible, use reference gas standards for control and calibration.

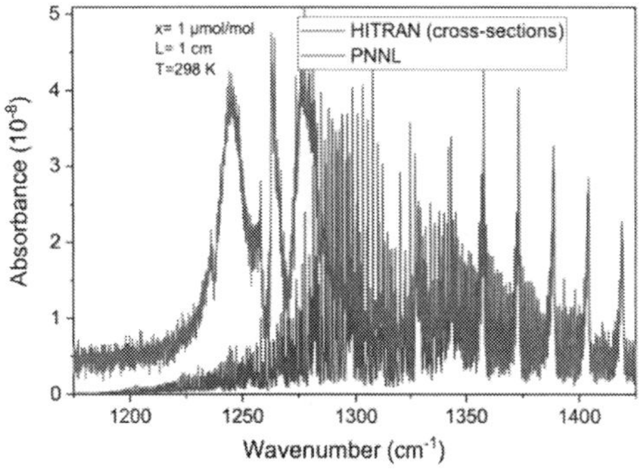

Figure 2 NO_2 absorbance based on PNNL and HITRAN database. In most spectral ranges there is good agreement but around 1263 cm^{-1} the PNNL database is too high due to the presence of an impurity (in this case HNO_2).

4 Is HNO_2 present in NO_2 gas standards?

In the pioneering work of Fried and co-workers [10] the presence of HNO_3 in NO_2 gas standards was demonstrated using the spectroscopic methods TDLS and FTIR, as well as by using chemiluminescence in combination with a nylon filter removing all acid gases like HNO_3. In 6 out of 7 cylinders, HNO_3 amount fractions of 2.2 to 23% relative to the NO_2 amount fraction were found. No HNO_2 was observed in these gas standards.

In literature, the formation of HNO_3 is typically given by:

$$2\ NO_2 + H_2O + \text{surface} \rightarrow HNO_3 + HNO_2 \qquad (1)$$

Based on this reaction, HNO_2 is expected to be present at similar amount fractions as HNO_3. While HNO_3 is consistently observed in NO_2 standards, HNO_2 has only rarely been reported in NO_2 gas standards [11]. Several laboratories have, in rare cases, observed HNO_2 but have no clear picture under which conditions HNO_2 is present (personal communication). In some cases, HNO_2 was only observed in one measurement of a mixtures but not in subsequent measurements of the

https://doi.org/(0.(05(/epjconf/10163510100(

same mixtures (e.g., BIPM analysis of one of the NPL mixtures in the CCQM K74:2018).

4.1 HNO_2 in pure NO

NO_2 gas standards are normally prepared starting from pure NO and not from pure NO_2 as the latter is only available at relatively low purity levels (~99.5%) compared to pure NO (99.99%). In high-purity nitric oxide, HNO_2 is observed at amount fractions of tens up to hundreds of µmmol/mol (Figure 3). Other impurities that are present include N_2O (increases in time in pure NO), HNO_3, NO_2, N_2O_3, CO_2, and H_2O.

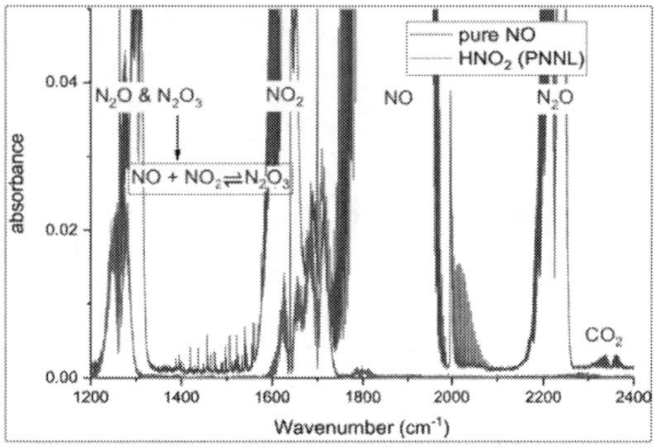

Figure 3 Measurement of pure nitric oxide showing the presence of N_2O, CO_2, N_2O_3 (exists in a dynamic equilibrium with NO and NO_2), and HNO_2. HNO_2 is typically present at around 50-300 µmol/mol in pure NO.

In diluted NO, N_2O_3 is no longer observed, as it is proportional to the product of the NO and NO_2 amount fractions.

4.2 HNO_2 absent in regular VSL NO_2 gas standards

In the VSL binary NO_2 gas standards, HNO_2 is not observed (Figure 4) in contrast to HNO_3. The HNO_3 amount fraction is proportional to the amount fractions of NO_2 and H_2O and is typically 1% relative to the NO_2 amount fraction.

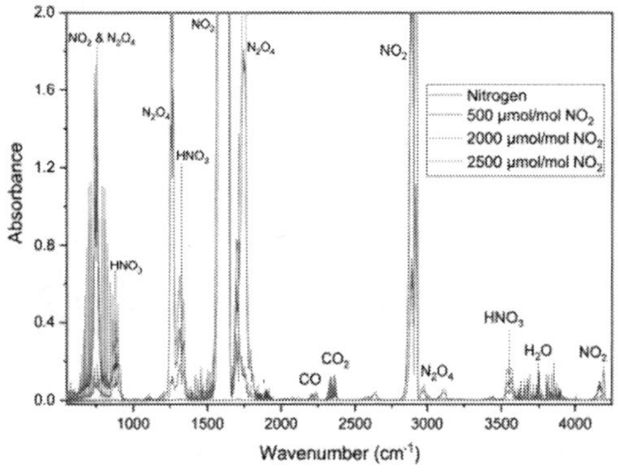

Figure 4 Measurement of high amount fraction NO_2 in N_2 standards showing the presence of the dimer (N_2O_4), HNO_3, H_2O, CO and CO_2. For the 2500 µmol/mol NO_2 mixture, the measured N_2O_4 amount fraction is 40 µmol/mol.

4.3 HNO_2 in PEMS gas standards

At VSL, HNO_2 was consistently observed for the first time in NO_2 gas standards that were produced for the Portable Emission Measurement Systems (PEMS). PEMS are used to determine the emission of road vehicles under real-driving conditions [12]. These PEMS gas standards contain a mixture of NO_2 and NO in a matrix with percentage levels of CO and CO_2 in nitrogen. Figure 5 shows an example of an FTIR measurement and the comparison with the PNNL database.

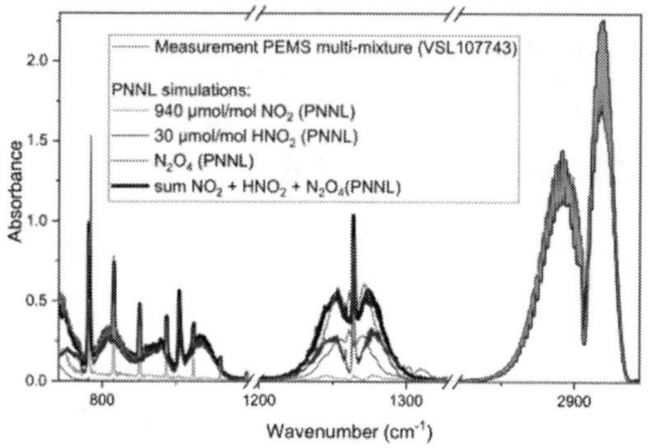

Figure 5 FTIR analysis of a PEMS multi-mixture (1001 µmol/mol NO_2, 2370 µmol/mol NO, 18% CO_2, 4.5 % CO in nitrogen). The picture shows 3 parts of the spectrum in which the FTIR measurement (—) is compared to a PNNL simulation (—) which is the sum of the individual NO_2, HNO_2 and N_2O_4 contributions.

In a set of 4 analysed PEMS mixtures, all mixtures contained HNO_2, while no HNO_3 was observed.

4.4 HNO_2 in CCUS gas standards

For CCUS there are several standards that set limits for impurities that can be present in the CO_2, as they can, for instance, have a negative impact on equipment or pipelines during transport or storage [13]. VSL prepares static reference gas standards of impurities in CO_2 to test and calibrate analysers for CCUS applications.

Figure 6 shows mixtures that have been prepared gravimetrically in a CO_2 matrix (left side) and the resulting analysis (right side). These mixtures were prepared to contain, next to NO_2, two sulphur components, water, and oxygen, and the matrix gas was CO_2. FTIR analysis revealed that NO_2 was significantly below the gravimetric amount fraction, while SO_2 was significantly higher. HNO_2 (a few percent relative to the NO_2 amount fraction) and NO were both formed, while HNO_3 was absent. Note that it is somewhat surprising that NO is present while there is also O_2 in these mixtures.

Possible reactions which might cause this change in amount fractions include:

$$H_2S + 3\ NO_2 \rightarrow SO_2 + H_2O + 3\ NO \quad (2)$$

$$NO_2 + NO + H_2O \rightarrow 2\ HNO_2 \quad (3)$$

https://doi.org/(0.(05(/epjconf/10163510100(

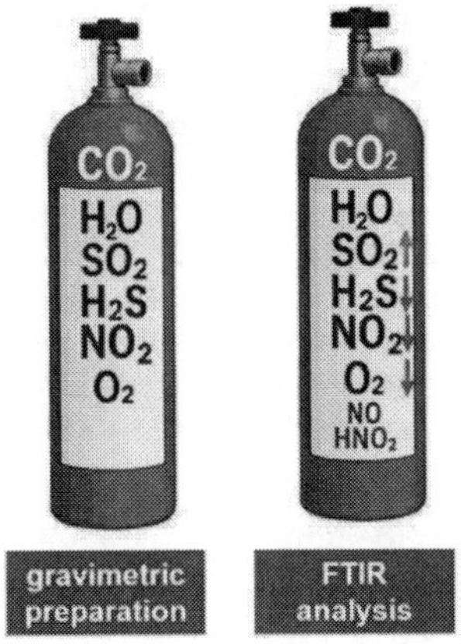

Figure 6 Gas standards with impurities at µmol/mol level in pure CO_2 were prepared using gravimetry (left cylinder). After preparation, the impurities were measured by FTIR except for H_2S and O_2 that were analysed with GC-SCD and GC-PDD. The analysis showed (right cylinder) that the SO_2 amount fraction was higher than the gravimetric amount fraction, while for H_2S, NO_2 and O_2 this was lower. Further, two additional components were formed in the standard: NO and HNO_2.

4.5 HNO_2 in NO_2-NO gas standards

From these PEMS and CCUS examples, it could be inferred that either the presence of CO_2 or the presence of NO is required for the formation of HNO_2 and the absence of HNO_3 in NO_2 gas mixtures. To test the latter hypothesis, 3 mixtures containing only NO and NO_2 in nitrogen were measured using CRDS. These measurements showed that in all 3 mixtures only HNO_2 is present while HNO_3 was not observed.

Table 2. Measured HNO_2 in three mixtures (10 µmol/mol NO + 10 µmol/mol NO_2 in N_2) of different ages.

Age mixture (years)	HNO_2 (relative to $NO_{2\,grav}$)	HNO_3
1	1.1%	Not detected
3	1.3%	
7	1.1%	

Based on the analysis results of the PEMS, CCUS and NO-NO_2 mixtures, it therefore seems that NO is a prerequisite for the presence of HNO_2 in NO_2 gas standards. This is confirmed by FTIR measurements of 6 NO_2 gas standards from a specialty gas company. Some of the mixtures contained also NO and these mixtures contained then indeed HNO_2 while the others did not.

A possible reaction leading to the formation of HNO_2 is:

$$NO + NO_2 + H_2O \rightleftharpoons 2\,HNO_2 \quad (4)$$

5 Conclusions and future outlook

The FTIR spectrometer is a solution for many of the analytical challenges for emerging applications like CCUS and hydrogen quality analysis. Using an optimised gas cell is important for correctly measuring reactive components. This paper provides the first systematic evidence that the presence of NO is a prerequisite for HNO_2 formation in NO_2 mixtures across multiple matrix environments.

In recent years, instruments based on mid-infrared supercontinuum lasers [14, 15] or frequency combs [16] have been developed which cover a large part of the wavelength range of a standard FTIR. Two of the main advantages of such laser-based instruments are the excellent beam quality, allowing the use of even more compact measurements cells (or even cavity enhancement to obtain kilometre long path lengths and down to 1 pmol/mol sensitivity for many molecules [17]!), and the high repetition rate, enabling fast measurements. In the next decade, such systems may take over the new role FTIR has just acquired in gas analysis.

Acknowledgements

Thanks to the colleagues from the Chemistry Group for preparing the gas standards and measurements. The research described in this paper received funding from the Ministry of Economic Affairs of the Netherlands.

ORCID iD

Stefan Persijn https://orcid.org/0000-0002-4356-4266

References

1. T. Stöhr, V. Reiter, S. Scheikl, N. Klopčič, S. Brandstätter, A. Trattner. Hydrogen quality in used natural gas pipelines: an experimental investigation of contaminants according to ISO 14687: 2019 standard. *International Journal of Hydrogen Energy* **67** (2024): 1136-1147. https://doi.org/10.1016/j.ijhydene.2023.09.305
2. K.R. Simonsen, D.S. Hansen, K. Fornitz, K. Kashefi, S Pedersen, S. (2025). Uncertainties in CO_2 Impurity Measurements Using FTIR. *IEEE Sensors Journal* 25(19) *36765-36772* https://doi.org/10.1109/JSEN.2025.3601622
3. L.P. Culleton, E.A. di Meane, M.K. Ward, V. Ferracci, S. Persijn, A. Holmqvist, K. Arrhenius, A. Murugan, P.J. Brewer, 2022. Characterization of Fourier transform infrared, cavity ring-down spectroscopy, and optical feedback cavity-enhanced absorption spectroscopy instruments for the analysis of ammonia in biogas and

biomethane. *Analytical chemistry*, **94** (44), pp.15207-15214 https://doi.org/10.1021/acs.analchem.2c01951

4. Thermo Fisher Scientific – Materials & Structural Analysis. (2024, May 16). Advancements in FTIR Gas Analyzers. AZoM. Retrieved on January 15, 2026 https://www.azom.com/article.aspx?ArticleID=22230
5. L.S. Rothman, (2021). History of the HITRAN Database. *Nature Reviews Physics*, *3*(5), 302-304 https://doi.org/10.1038/s42254-021-00309-2
6. K. Bielska, A.A. Kyuberis, Z.D. Reed, G. Li, A. Cygan, R. Ciuryło, E.N. Adkins, L. Lodi, N.F. Zobov, V. Ebert and D. Lisak, 2022. Subpromille measurements and calculations of CO (3–0) overtone line intensities. *Physical Review Letters*, **129** (4), p.043002 https://doi.org/10.1103/PhysRevLett.129.043002
7. https://www.bipm.org/en/committees/cc/ccqm/wg/ccqm-gawg-tg-adv-spec
8. R.V. Kochanov, I.E. Gordon, L.S. Rothman, S.W. Sharpe, T.J. Johnson, R.L. Sams (2015). Comment on" Radiative forcings for 28 potential Archean greenhouse gases" by Byrne and Goldblatt (2014). *Climate of the Past*, **11** (8), 1097-1105 https://doi.org/10.5194/cp-11-1097-2015
9. S. Persijn, F. Harren, A van der Veen, A. (2010). Quantitative gas measurements using a versatile OPO-based cavity ringdown spectrometer and the comparison with spectroscopic databases. *Applied Physics B*, **100** (2), 383-390 https://doi.org/10.1007/s00340-009-3875-3
10. A. Fried, A., R. Sams, W. Dorko, J.W. Elkins, Z. Cai, Z. Determination of nitrogen dioxide in air-compressed gas mixtures by quantitative tunable diode laser absorption spectrometry and chemiluminescence detection. Analytical Chemistry, **60**(5), 394-403 (1988) https://doi.org/10.1021/ac00156a005
11. Final publishable report, $MetNO_2$ project, (2020) http://empir.npl.co.uk/metno2/wp-content/uploads/sites/23/2021/04/MetNO2_Final_Publishable_Report_v7-ACCEPTED.pdf
12. S. Persijn, E. Efremov, S. Lombard and M. Carré D2: Best practice guide on the use of high amount fraction i.e., up to at least 2500 µmol/mol NO_2 reference materials, MetroPEMS project, 30 April, 2023
13. ISO 27913:2024 Carbon dioxide capture, transportation and geological storage — Pipeline transportation systems, International Organization for Standardization, Geneva, Switzerland, 2024.
14. I. Zorin, P. Gattinger, A. Ebner, M. Brandstetter, M. (2022). Advances in mid-infrared spectroscopy enabled by supercontinuum laser sources. *Optics Express*, **30** (4), 5222-5254 https://doi.org/10.1364/OE.447269
15. R. Krebbers, K. van Kempen, F.J.M. Harren, S. Vasilyev, I.F. Peterse, S. Lücker, A. Khodabakhsh, S.M. Cristescu (2024) Ultra-broadband spectroscopy using a 2–11.5 µm IDFG-based supercontinuum source. *Optics Express* **32** (8): 14506-14520 https://doi.org/10.1364/OE.515914
16. H. Timmers, A. Kowligy, A. Lind, F.C. Cruz, N. Nader, M. Silfies, G. Ycas. T.K. Allison, P.G. Schunemann, S.B. Papp, S.A. Diddams (2018) Molecular fingerprinting with bright, broadband infrared frequency combs. *Optica* **5** (6): 727-732 https://doi.org/10.1364/OPTICA.5.000727
17. Q. Liang, A. Bisht, A. Scheck, P Schunemann, J. Ye (2025). Modulated ringdown comb interferometry for sensing of highly complex gases. *Nature*, 1-8 https://doi.org/10.1038/s41586-024-08534-2

) Ht sp:/ odor. :g0ig05g:et jconf:101235101001

Gaseous reference standard for breath analysis

Laura Gagni [1*], *Giorgio Bissolotti*,[1]

[1] SIAD S.p.A., Gases of life District, Osio Sopra, Italia

Abstract. An increasing amount of research is going into detecting volatile organic compounds (VOCs) in exhaled breath to detect diseases and personalize treatment for patients: recently, these analytes has risen as potential biomarkers. The present work aims to examine different multi-component gas mixtures suitable for breath gas analysis. In addition, also the stability of these mixtures was investigated. The number of components tested were 31, divided in two gas mixtures A and B. These components are aldehydes, acetate compounds, aromatics compounds: the lists are available in the article. Gas components of mixture A and B were prepared in different concentration from 1 µmol/mol to 10 nmol/mol. All the mixtures were analytically verified and the expected analytical relative uncertainty is 10-15% for 10 nmol/mol mixture. The stability check is also performed after 1 year and 2 years. The mixtures demonstrated good stability, and tests are still on going.

1 Introduction

Research into volatile organic compounds (VOCs) in exhaled breath has grown substantially in recent years, as these analytes have emerged as valuable potential biomarkers for disease detection and personalized treatment. Once collected and analysed, VOCs can provide a rapid, accurate, non-invasive, and painless diagnostic tool for a wide range of diseases and health conditions.

However, breath samples are inherently challenging to collect because they are gaseous and can contain high levels of noise due to VOCs present in inhaled ambient air. For analytical purposes, gas chromatography–mass spectrometry (GC-MS) is widely recognized as the gold standard, as it enables the detection of chemically diverse compounds within complex mixtures with high sensitivity and across a broad dynamic range. GC-MS also allows accurate quantification of VOCs and provides high confidence in biomarker identification [1].

To ensure reliable results, the use of standards for calibrating detection systems is essential. This guarantees consistency across devices and prevents analytical drift over time. In practice, liquid-phase standards are more commonly used because they offer several advantages, including low cost, ease of handling, and suitability for a broad range of VOCs. They also allow the application of internal standard methods [2,3]. Nevertheless, research has shown that calibrating gaseous samples with liquid-phase standards can introduce substantial variability and measurement bias [4]. For this reason, gas-phase standards are often preferred for VOC analysis, as they help eliminate or reduce matrix effects by ensuring that both the sample and the calibration standard are in the same phase. Furthermore, gas-phase standard mixtures can be formulated to include all VOCs of interest, including isotopically labelled analytes at very low concentrations. The present work focuses on the preparation and analysis of multi-component gas mixtures suitable for breath analysis, with concentrations ranging from 1 µmol/mol down to 10 nmol/mol. The stability of these mixtures was also investigated. In total, 31 compounds were examined, divided into two gas mixtures. These compounds can be combined with other commonly used VOCs whose stability is already well established, even at very low concentrations.

2 Material and methods

2.1 Preparation of static standards

A total of 31 target compounds were selected and divided into two gas mixtures, A and B, to ensure their stability and to avoid incompatibilities among certain components. The compositions of the two mixtures are reported in Table 1 and Table 2. The producer prepared the mixtures starting from pure substances with the highest commercially available purity. All chemicals were GC-grade analytical standards from Sigma Aldrich, with stated purities of 98% or higher. Nitrogen was used as the balance gas, with a purity of 99.9995% mol, supplied by SIAD S.p.A.

The mixtures were filled into 10-litre aluminium cylinders made of 6061 alloy, supplied by Luxfer, and fitted with brass valves. Prior to filling, both the cylinders and the valves underwent a dedicated conditioning

* Corresponding author: laura_gagni@siad.eu

procedure designed to remove residual moisture and other impurities. Particular attention was given to the presence of water, both in the nitrogen used as balance gas and throughout the filling procedure. Following a gravimetric method inspired to ISO 6142-1, the mixtures were prepared by adding the components sequentially into the cylinders and weighing each addition using calibrated laboratory balances (Sartorius CCE60K2, Sartorius MAS5203S-100DR, and Sartorius MSA524S-100-DU)

Table 1. list of components in mixture A and relative CAS-Number

Mixture A		
N	Component	CAS Number
1	Styrene	100-42-5
2	Toluene	108-88-3
3	Deuterated Toluene	2037-26-5
4	(R)-Limonene	5989-27-5
5	Isoamyl Acetate	123-92-2
6	Perfluoro-2-methyl-3-pentanone	756-13-8
7	Acrolein	107-02-8
8	Acrylonitrile	107-13-1
9	Benzene	71-43-2
10	Deuterated Benzene	1076-43-3
11	Butyl Acetate	123-86-4
12	Cyclohexene	110-83-8
13	1,1-Dichloroethylene	75-35-4
14	Diethyl Ether	60-29-7
15	Ethyl Acetate	141-78-6
16	Fluorobenzene	462-06-6
17	1-Heptene	592-76-7
18	1- Hexene	592-41-6
19	Isoprene	78-79-5
20	Methyl Acetate	79-20-9
21	1- Octene	111-66-0
22	Propyl Acetate	109-60-4

Table 2. list of components in mixture B and relative CAS-Number

Mixture B		
N	Component	CAS Number
1	Valeraldehyde	110-62-3
2	Hexanal	66-25-1
3	Benzaldehyde	100-52-7
4	Octanal	124-13-0
5	Heptanal	111-71-7
6	Decanal	112-31-2
7	Nonanal	124-19-6
8	Isobutyraldehyde	78-84-2
9	Propionaldehyde	123-38-6

2.2 Production scheme

Gas mixtures A0 and B0 were prepared in two cylinders at a concentration of 2 µmol/mol. Starting from these two primary mixtures, additional diluted gas mixtures containing the same components were produced with gravimetric method. All mixtures were prepared in the cylinders described in paragraph 2.1, and at different times, in order to ensure process reliability and to enable stability testing. Further details are provided in Table 3.

Table 3. scheme of production of mixtures started from mixture A.

Mix code	Conc. Mix A in nitrogen	Mix code	Conc. final mixture in nitrogen
A0	2 µmol/mol		
A0 + N2	=	A1, A4	1 µmol/mol
A0 + N2	=	A5	100 nmol/mol
A0 + N2	=	A2	50 nmol/mol
A0 + N2	=	A3	10 nmol/mol

Mix code	Conc. Mix B in nitrogen	Mix code	Conc. final mixture in nitrogen
B0	2 µmol/mol		
B0 + N2	=	B1, B4	1 µmol/mol
B0 + N2	=	B5	100 nmol/mol
B0 + N2	=	B2	50 nmol/mol
B0 + N2	=	B3	10 nmol/mol

As mentioned earlier, the preparation of these gas mixtures was carried out at two different times. Mixtures A4, A5, B4, and B5 were prepared in August 2023, while all the others were produced in April 2023.

2.3 Analysis of gas mixture

All mixtures were analytically verified, and all results were positive. The instrument was calibrated using standards certified by SIAD S.p.A. Each mixture was also subjected to cross-verification, both against the other mixtures and against their parent mixtures A0 and B0. A GC-FID system (Agilent Technologies) was used to quantify all components. Separation was achieved using a Vocol column (Supelco), and the oven temperature was programmed from 40 °C to 200 °C to ensure adequate resolution of the analytes. Special attention was given to the sampling procedure to guarantee that a representative amount of gas reached the GC, thereby preventing losses of high-molecular-weight compounds. All components exhibited good chromatographic resolution, with the exception of diethyl ether and isoprene, which co-eluted due to identical retention times.

2.4 Stability tests

The stability assessment was carried out using mixtures A4, A5, B4, and B5. These mixtures were analysed following the methodology described in paragraph 2.3. They were certified in August 2023 and subsequently analysed in April 2024 (corresponding to 8 months of stability) and again in January 2026

(29 months of stability). All mixtures were verified through cross-comparison, pairing A4 with A5 and B4 with B5.

3 Result and discussion

3.1 Preparation results

The gas mixtures produced at different concentration levels have a target concentration, a gravimetric concentration and an analysed concentration. For all the mixtures and components, the agreement was very good with differences below 15 % as shown in Fig. 1 and Fig. 2. The difference between analytical value (x_a) at the time of production and the gravimetric value (x_g) for each component i is represented by the evaluation of the parameter D_i:

$$D_i = (x_g - x_a)/x_g \% \quad (1)$$

Fig. 3 shows the results for mixture A4. The values of D_i exhibit an acceptable and fairly symmetrical distribution around zero: twelve components have D_i values below zero, while nine components show values above zero. Fig. 4 presents the results for mixture B5. In this case, the distribution is more unbalanced: six components display analytical concentrations significantly lower than their gravimetric values. The remaining components — propionaldehyde, isobutyraldehyde, valeraldehyde, and caproaldehyde— show good agreement between gravimetric and analytical concentrations.

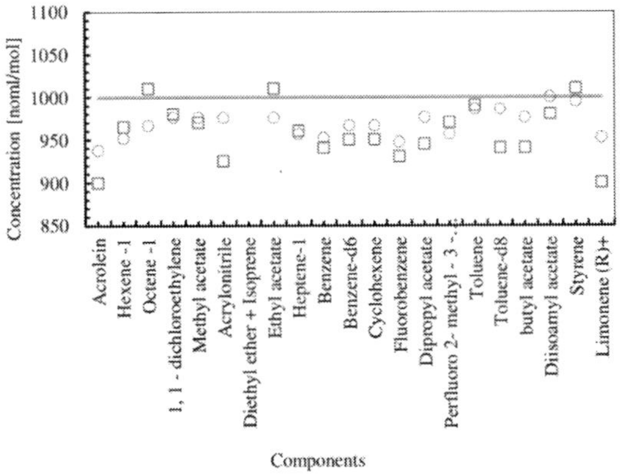

Fig. 1. For mixture A4 comparison of target concentration (line), gravimetric concentration (symbol □) and analytical concentration (symbol ○).

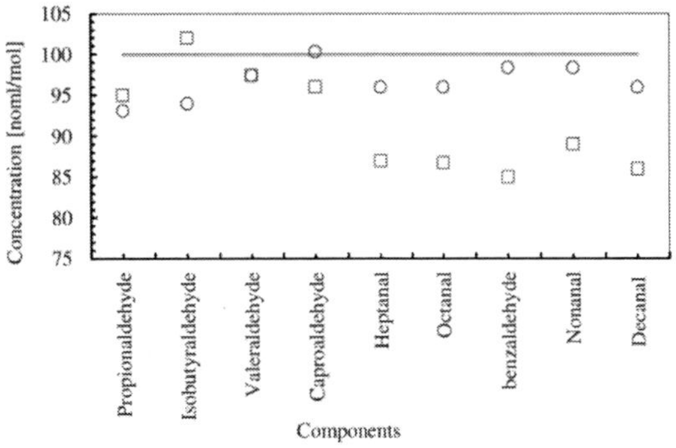

Fig. 2. For mixture B5 comparison of target concentration (line), gravimetric concentration (symbol □) and analytical concentration (symbol ○).

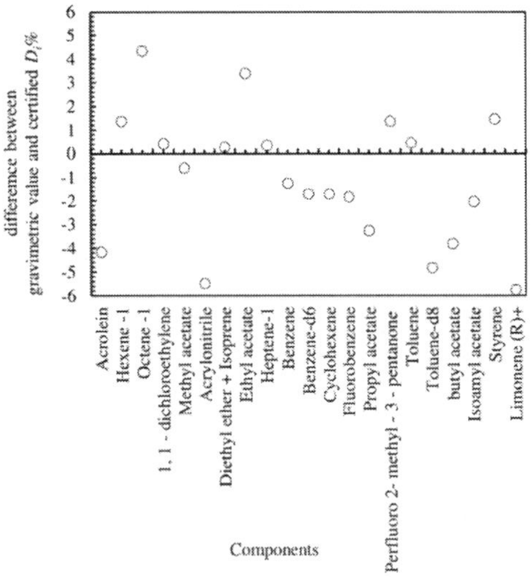

Fig. 3. For mixture A4 parameter D_i (symbol ○).

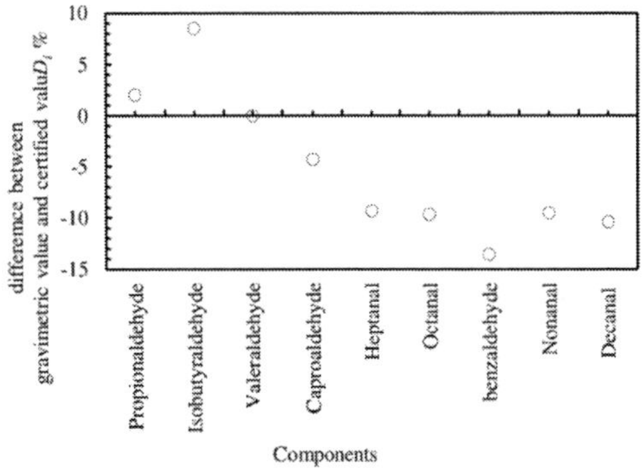

Fig. 4. For mixture B5 parameter D_i (symbol ○).

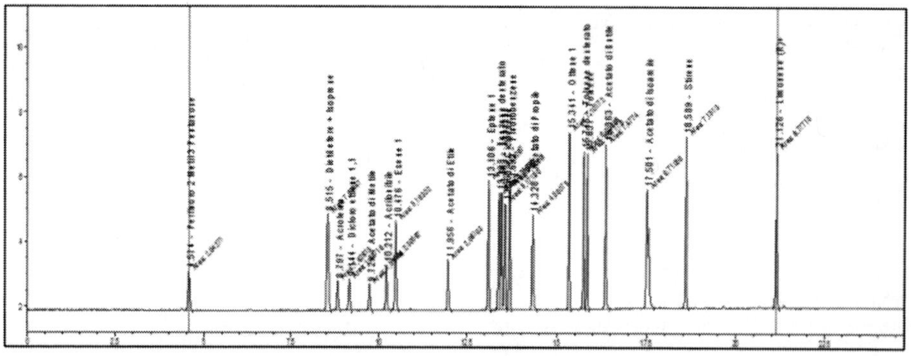

Fig. 5 GC- FID chromatogram for mixture A4.

All components showed good chromatographic resolution as shown in Fig. 5 for mixture A4 and Fig. 6 for mixture B5.

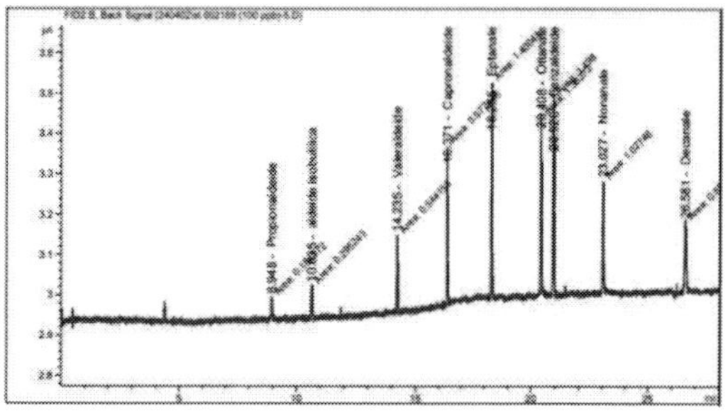

Fig. 6. GC- FID chromatogram for mixture B5.

3.2 Stability results

All the mixtures were verified with a cross control between them A4 with A5 and B4 with B5. This approach could be possible thanks to one of the most important properties of the FID : its extreme long linear range. The linearity has its limitations according to studies dealing with the linear behaviour of the detector response [5–7]. The mixture A4 was used to calibrate the instrument and analyse mixture A5. Fig 7. show the results.

All the components have a concentration of approximately 100 nmol/mol. The analysis performed in August 2023, April 2024 and January 2026 show excellent agreement for all components except ethyl acetate, propyl acetate and butyl acetate. For these components the consistency is observed only between the two later analyses (April 2024 and January 2026), while the certified value from August 2023 is lower than the values obtained during the stability assessments.

The results confirm the stability of great part of components in the mixtures which maintain their stability for more than 2 years at both levels of concentration. However, further investigation is required to better understand the behaviour of acetate compounds.

The mixture B5 was used to calibrate the instrument and analyse mixture B4. Fig 8. show the results.
The data indicate good stability for isobutylaldehyde and propionaldehyde while some critical issues emerge for high molecular weight aldehydes. In particular:

- for valeraldehyde, capraldehyde and benzaldehyde agreement is observed only

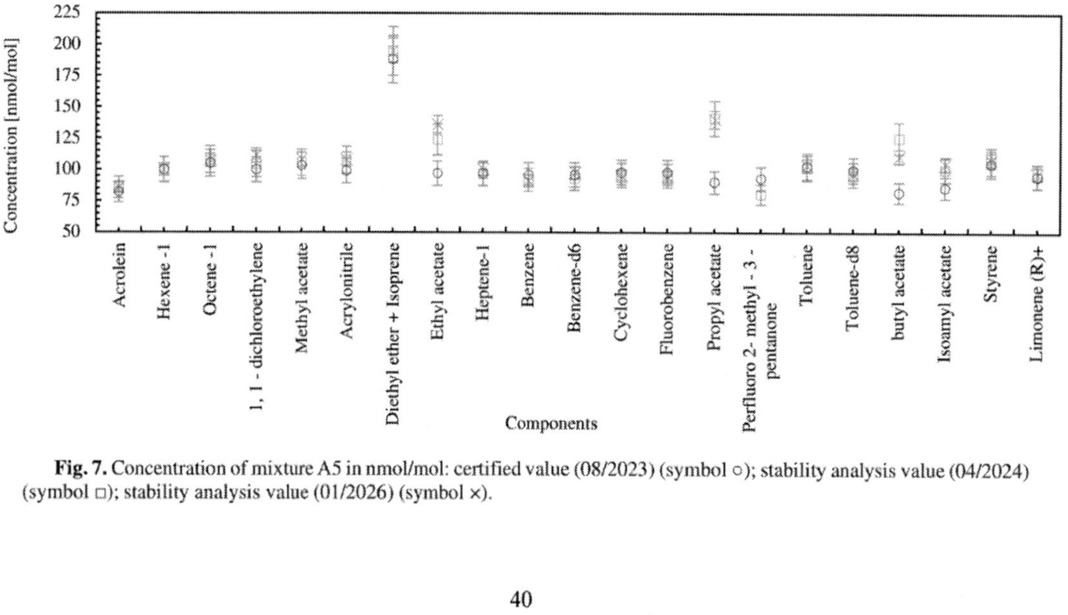

Fig. 7. Concentration of mixture A5 in nmol/mol: certified value (08/2023) (symbol ○); stability analysis value (04/2024) (symbol □); stability analysis value (01/2026) (symbol ×).

between the most two recent analysis (April 2024 and January 2026) whereas the certified value from August 2023 is higher than the stability measurements.

- Heptanal, octanal, nonanal and decanal exhibit a continuous decrease in concentration over time, indicating that these components are not stable.

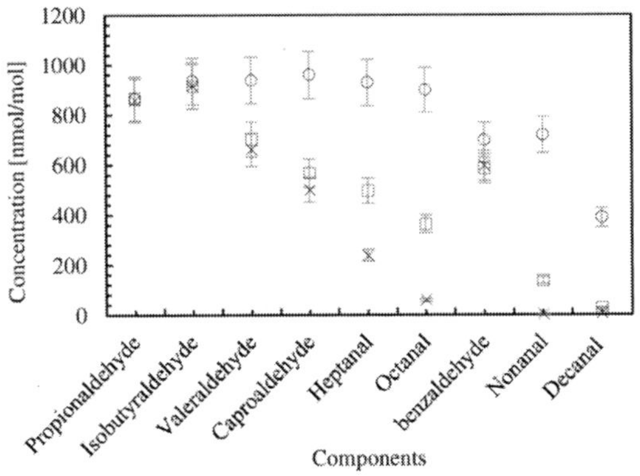

Fig. 8. Concentration of mixture B4 in nmol/mol: certified value (08/2023) (symbol ○); stability analysis value (04/2024) (symbol □); stability analysis value (01/2026) (symbol ×).

These findings are consistent with the results obtained during mixture preparation. The parameter $D_i$for the higher-molecular-weight aldehydes (heptanal, octanal, nonanal, and decanal) was approximately −10% or lower, suggesting that these compounds may undergo partial adsorption or degradation even shortly after preparation. This behaviour likely contributes to the observed instability during long-term storage. That indicates different possible scenario:

1. The theoretically weighed amount of product did not fully enter the cylinder and was partially lost during the filling operations.
2. The cylinder was correctly filled and the entire weighed amount was transferred, but the sampling procedure did not allow the full quantity to reach the GC-FID detector, resulting in an apparently lower concentration.
3. The components inside the cylinders were not stable, and their concentrations gradually decreased over time.

Further studies are required to better understand the behaviour of high-molecular-weight linear aldehydes, considering additional variables such as different concentration levels during stability testing..

4 Conclusion

The present study demonstrates the feasibility of producing stable calibration gas mixtures for breath analysis. The mixtures developed include aldehydes, acetate compounds, aromatic compounds, alkenes, and deuterated species—representative of the key VOC classes commonly investigated in breath research. Very low concentration levels can be achieved, with stability maintained for more than two years. Further work will focus on elucidating the behaviour of acetate compounds and higher-molecular-weight aldehydes, for which specific issues were observed.

Additionally, deeper interpretation may be obtained through supplementary analysis of the raw chromatographic data, taking advantage of the proportional relationship between FID signal intensity and the carbon content of hydrocarbons

References

1. White paper, Breath Biopsy® OMNI®: Advanced Global Breath VOC Analysis.
2. M.A. Iqbal, K.H. Kim, J.E. Szulejko, M.M. Rahman The selection of the standard phase (gas vs. liquid) and the related matrix effect on the direct injection gas chromatographic analysis of VOCs at sub–ppm levels. Atmos. Pollut. Res., 5 (3) (2014), pp. 563-571, 10.5094/APR.2014.065
3. C. Liaud, N.T. Nguyen, R. Nasreddine, S. Le Calvé. Experimental performances study of a transportable GC-PID and two thermo-desorption based methods coupled to FID and MS detection to assess BTEX exposure at sub-ppb level in air.

   Talanta, 127 (2014), pp. 33-42, 10.1016/j.talanta.2014.04.001 J.
4. Y.H. Kim, K.H. Kim: Effect of standard phase differences between gas and liquid and the resulting experimental bias in the analysis of gaseous volatile organic compounds. Anal. Chim. Acta, 714 (2012), pp. 98-103, 10.1016/j.aca.2011.11.042
5. McWilliam, I. G. "A study of the flame ionization detector", Journal of Chromatography A, 51, pp. 391–406, 1970. https://doi.org/10.1016/S0021-9673(01)96888-8
6. Slemr, J., Slemr, F., D'Souza, H., Partridge, R. "Study of the relative response factors of various gas chromatograph–flame ionisation detector systems for measurement of C2 –C9 hydrocarbons in air", Journal of Chromatography A, 1061(1), pp. 75–84, 2004. https://doi.org/10.1016/j.chroma.2004.10.037
7. Bromly, J. H., Roga, P. "Nonlinearities in Flame Ionization Detector Response", Journal of Chromatographic Science, 18(11), pp. 606–613, 1980 https://doi.org/10.1093/chromsci/18.11.60

https://doi.org/50.50j 5/ep3conf/1016(j 10100(

Data-driven models for the operation of a dynamic primary standard based on permeation

Federica Gugole[1], *Adriaan M. H.* van der Veen [1,2*], and Ewoud J.J. de Boed[2]

[1]VSL, Data Science & Modelling Department, Thijsseweg 11, 2629 JA Delft, The Netherlands
[2]VSL, Chemistry Department, Thijsseweg 11, 2629 JA Delft, The Netherlands

Abstract. For the preparation of calibration gas mixtures with low levels of reactive compounds, the permeation method as described in ISO 6145-10 is excellently suited. In the past years, it has proved to be useful for developing primary standards for key impurities in hydrogen, such as ammonia, hydrogen fluoride and hydrogen chloride. Using an automated balance to measure the mass loss of the permeation tube, large volumes of data are gathered, from which the permeation mass flow rate is calculated. The residuals of traditional simple straight-line regression show patterns which deserve further attention. They suggest that the current approach may underrate the uncertainty associated with the permeation rate. The permeation system is operated with a dilution system using thermal mass flow controllers. These produce large volumes of data as well, which are processed to obtain a mass flow rate and associated uncertainty. It is shown how time series models enable assessing correlations in the data from the magnetic suspension balance and the thermal mass flow controllers. These data-driven models provide a better description of the features of the data and thereby provide more realistic estimates of the mass flow rates and associated uncertainties. The time series analysis reveals that the permeation data are quite heavily autocorrelated, whereas the data from the thermal mass flow controllers are uncorrelated.

1 Introduction

Permeation is one of the measurement principles used by the gas industry and National Metrology Institutes (NMIs) to create calibration gas mixtures. The method is described in ISO 6145-10 [1] and often applied in conjunction with using thermal mass flow controllers (MFCs) [2] to dose the matrix gas. The method enables preparing calibration gas mixtures with the component of interest at μmol mol^{-1}-level in one step. The method can be used for a wide range of components including sulfur dioxide, nitrogen dioxide [3], hydrogen chloride, hydrogen fluoride and formaldehyde [4]. The method was successfully applied in key comparisons [5, 6] and for developing a primary measurement standard for key impurities in hydrogen [4, 7], among many other applications.

Especially the use of a magnetic suspension balance (MSB) to continuously weigh the permeation tube enables high-end applications. The MSB is used to record the mass loss as a function of time. The data are used as the mass flow rate of the component of interest. The permeation tube is filled with a gas, liquid or solid of adequate purity that permeates through the wall. The substance that permeates through the wall is diluted with the matrix gas(es) and carried to the analyser. A typical data set from the permeation of hydrogen chloride is shown in figure 1. The slope of the fitted line is the permeation rate of the hydrogen chloride.

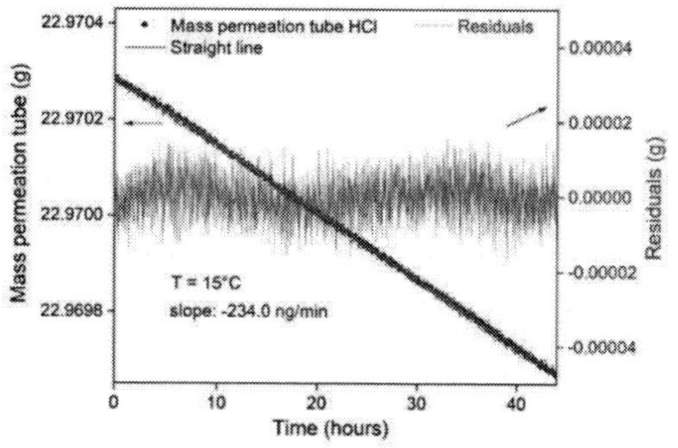

Fig. 1. Permeation data of hydrogen chloride at 15 °C. The results (fitted straight line and residuals) of OLS regression are also displayed.

The patterns in the residuals from fitting a straight line through the permeation data reveal that they are not independent and identically distributed (IID). These patterns hint at the presence of autocorrelation in the data. Autocorrelation means that a datum in a time series depends on its predecessor or a few predecessors [8, 9]. Just as with other kinds of dependencies, the presence of autocorrelation in a data set influences any results calculated from it.

* Corresponding author: avdveen@vsl.nl

An MSB generates a lot of data, equally spaced in time, which lends itself well for the application of time series analysis. In this work, time series analysis is applied to (1) assess the structure of the data and (2) specify a model that describes the dependencies in the data. Then, the time series analysis is combined with the regression to replace the ordinary least squares (OLS) commonly used when obtaining the permeation rate.

2 Analysis

2.1 Time series

Differently from a more general data set, in a time series the (temporal) order in which the data have been recorded matters. Time series models thus appreciate the temporal relation between subsequent data points and enable modelling phenomena with a temporal behaviour and capture any resulting correlations. More classical statistical techniques such as OLS instead, are not sensitive to the order in which the data are stored in the data set, and therefore they cannot capture any temporal dependency among the data.

Measurement data collected using a dynamic method, such as permeation, naturally have a temporal component and the order in which these data are collected matters since it gives information on the behaviour of the system. Therefore, data sets collected by such systems are natural candidates for time series analysis.

2.2 MSB data set

The first data set analysed in this work is the data set from the permeation of hydrogen chloride shown in figure 1. The data set covers more than 40 hours of measurements recorded on average every 0.67 min. The residuals of the OLS regression displayed in figure 1 (blue line) show an oscillating behaviour which hints that they are not IID. Factors that may contribute in this regard are the temperature control and the pressure regulator. The permeation chamber is temperature and pressure controlled, thus the heater switches on and off when the temperature reaches predefined thresholds, and a similar behaviour is expected also by the pressure regulator.

The MSB performs an automatic zeroing after every sixth measurement. Therefore, the time interval between the sixth and seventh measurement is about 2.23 min. Since time series analysis requires the data to be equally spaced in time, the question arises whether these gaps should be filled in the series to have a more homogeneous temporal spacing over the whole series. However, this would require imputing two data points in each auto-zeroing interval. Consequently, imputing would lead to 25 % of the final time series to be imputed data with statistical properties potentially different with respect to the original data. Given the large percentage of imputed data, the results of the analysis might be significantly affected by the chosen imputing method. Therefore, it was decided to analyse the dataset as is and to not impute any data.

2.3 MFC data set

The second data set considered here is a data set collected during a stability test of the MFCs used in the permeation standard. There, an MFC with 100 mL min^{-1} capacity was set to a volume flow rate of 50 mL min^{-1}. The actual volume flow rate was recorded from the MFC sensor and from a piston prover. The test lasted about twenty hours and data were recorded every movement of the piston and then averaged per 10 measurements. The data set containing the MFC readings is shown in figure 2.

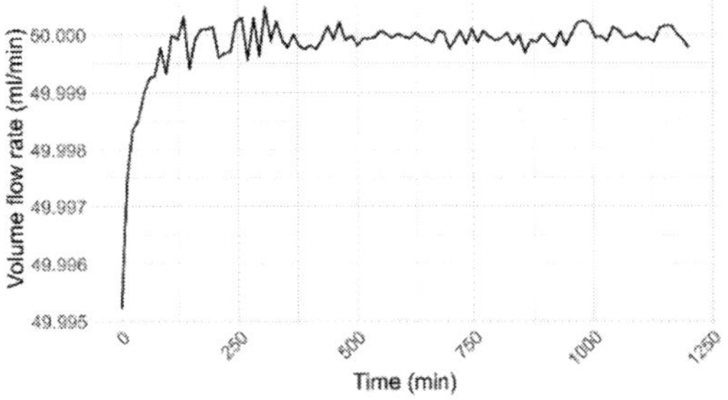

Fig. 2. Observed volume flow rate recorded by a mass flow controller during a stability test.

After an initial warm-up of the MFC, the recorded flow rate oscillates around a constant value. In the time series analysis the initial warm-up phase is neglected as it is clear that in that phase the properties of the system are different from the remaining series of data, thus breaking the assumption of (weak) stationarity, see [8, 9] for more information. The warm-up and stabilization were confirmed by the measurements using the calibrated piston prover.

2.4 Time series model

When fitting the MSB dataset with OLS, the so-obtained residuals show some visible structure, see figure 1. This indicates that the assumptions for the OLS algorithm are not satisfied and that other statistical techniques should be used to perform the regression. In particular, if the assumption that the error terms are uncorrelated is broken, the OLS algorithm will tend to underestimate the standard errors associated to the parameters.

In this work, the so-called seasonal autoregressive integrated moving average (SARIMA) model [8, 9], a generalisation of the autoregressive moving average (ARMA) model to periodic variations, was applied to the residuals of the OLS regression following a procedure similar to the one described in [10, 11] for ARMA processes. SARIMA models allow to model both the seasonal and the within season behaviour of the data as an ARMA model, where the order of each model is to be determined as described in [10] and keeping in mind that only the autocorrelation function (ACF) and partial ACF (PACF) entries at multiples of the seasonality are to be considered when determining the model for the seasonal part. Lastly, a straight-line with autocorrelated residuals was fitted to the data and the result compared with the outcome of the OLS regression.

The MFC dataset does not display visible structure. Therefore, in this case time series analysis is used to check if there is any hidden pattern causing (auto-) correlation or if the assumption of IID values holds.

3 Results

3.1 Regression with autocorrelated residuals

Both the ACF and PACF of the OLS residuals show significant correlation every sixth data point, see figure 3. This coincides with the timings of the automatic zero autocalibration regularly performed by the MSB. Thus, it is reasonable to model this as a seasonal (or cyclic) event with a given seasonality length of six.

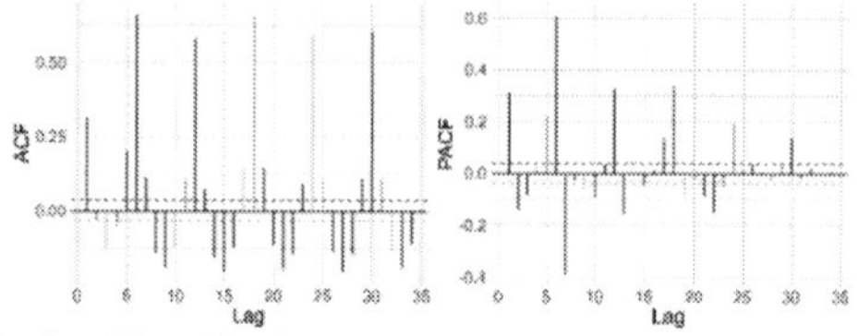

Fig. 3. ACF and PACF values of the residuals obtained fitting a straight line to the MSB data set using OLS.

Since the correlation peaks due to the seasonality are more or less all of the same value in the ACF but are decreasing in the PACF, it was decided to model the seasonality as an autoregressive (AR) model of order 1. This is not the only correlation present in the data though. Within the six-measurement cycle it is possible to note that measurements at distance one, i.e., measurements that are recorded subsequently after each other, are also correlated. The PACF displays less significant correlations within a single season, so an AR model of order 1 is used also for the within season model component.

The residuals of the resulting SARIMA model do not show any visible structure and their ACF and PACF show much less correlation, see figure 5. In particular, the peaks at regular intervals have disappeared, indicating the seasonal component has been correctly captured by the model. The oscillating pattern visible in the OLS residuals has also disappeared (see figure 4) thus it has also been captured by the SARIMA model. Both the ACF and PACF still show some significant correlations with higher terms (i.e., data points with higher temporal separation between each other). However, no clear pattern is visible and the correlation values are rather low (approximately 0.2 in absolute value), so it is concluded that this correlation is mostly due to noise and it is not necessary to further complicate the model.

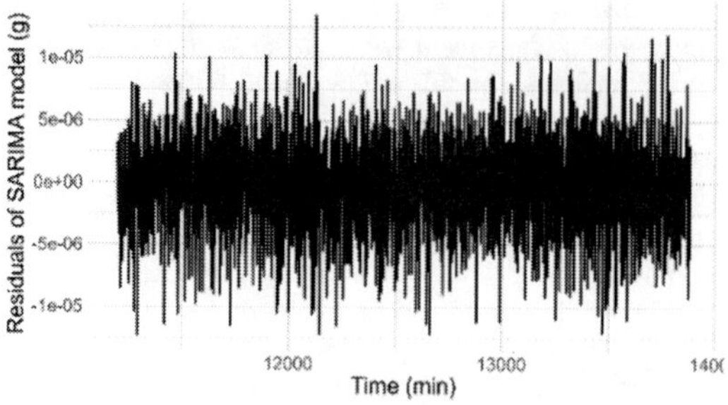

Fig. 4. Plot of the residuals of the SARIMA model applied to the OLS residuals.

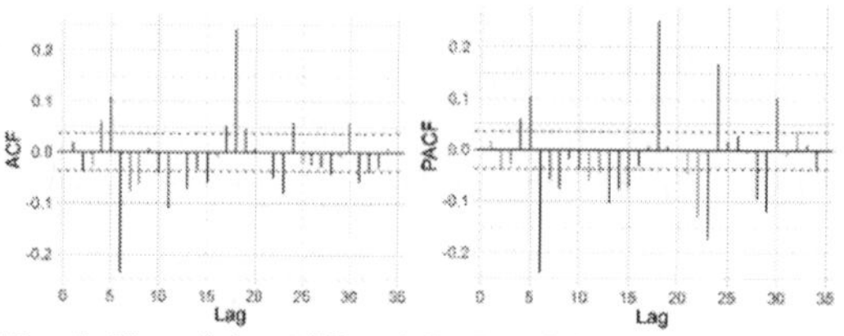

Fig. 5. Plot of the ACF and PACF of the residuals of the SARIMA model applied to the OLS residuals.

The SARIMA model for the residuals can thus be included in the regression algorithm. In table 1 the regression parameters and associated standard errors calculated with or without correlation among the OLS residuals are reported. The mean estimate of the parameter values does not change significantly but the standard error is almost four times larger when including the SARIMA model for the residuals.

Table 1. Regression coefficients value and associated standard error obtained either using OLS (i.e., assuming independence of the residuals) or including a SARIMA model for the residuals.

	Coefficient	**Value**	**Std. error**
OLS	Intercept	22.97	$1.64 \cdot 10^{-6}$
	Slope	$-2.340 \cdot 10^{-7}$	$1.31 \cdot 10^{-10}$
SARIMA	Intercept	22.97	$6.06 \cdot 10^{-6}$
	Slope	$-2.340 \cdot 10^{-7}$	$5.15 \cdot 10^{-10}$

3.2 Stability of MFC

Since no visible structure in the MFC data set is visible in figure 2, the ACF and PACF values (see figure 6) may be checked to ensure that the IID assumption can be used for this case.

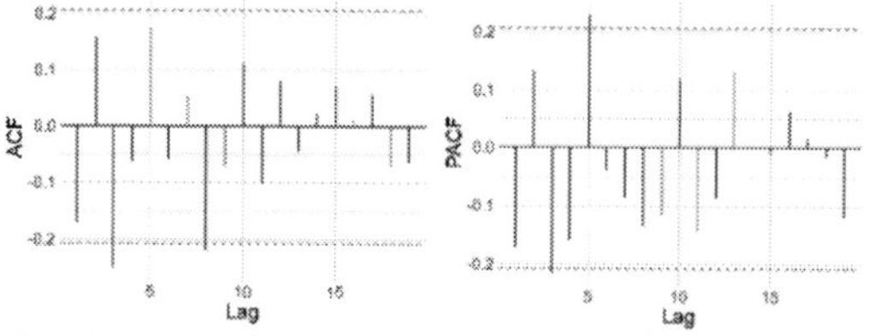

Fig. 6. ACF and PACF values of the MFC data recorded during a stability test.

Most of the ACF and PACF are not statistically different from zero since they are within the 95 % confidence interval defined by the blue lines. Those whose value exceeds the limit of the confidence interval are only barely outside, with a correlation value rather low (about 0.2) and they are at higher terms. It was thus concluded that there is indeed no underlying structure in the data, and the data can be treated as independent.

4 Conclusions

Dynamic methods for preparing calibration gas mixtures, such as permeation, generate data sets where the order in which the data have been recorded matters. Time series models thus are natural candidates for the analysis of such data sets since they can capture the temporal relationship between the measurement results and thus determine any potential correlation.

This work showed two practical examples of how time series analysis can effectively be used to estimate (auto-) correlations, if any, between measurements. In particular, their application to a data set from the permeation of hydrogen chloride recorded by a magnetic suspension balance allowed to incorporate the regular automatic zero-recalibration performed by the balance in the regression model estimating the permeation rate. This resulted in an almost four times larger standard error associated to the parameters estimated by the regression model with respect to the case where simple straight-line regression is used.

The application of the (partial) autocorrelation function to a data set collected by a mass flow controller during a stability test revealed that for such a test the assumption of IID measurement results is acceptable and may thus be used in the uncertainty analysis.

In both cases, the results of the time series analysis can be readily included in the uncertainty budget of the dynamic method. In the case of the analysis on the MSB data sets, the standard error associated with the permeation rate estimated by the time series model would have be used instead of the one obtained used OLS. In the case of the analysis on the data set recorded by the MFCs, no changes are required in the uncertainty budget since in that case the time series analysis simply confirmed that the assumption of IID values holds.

Time series analysis is therefore a useful tool in the analysis of data sets recorded by dynamic methods or systems. More work on how to combine and propagate measurement uncertainty in the presence of serial correlation is under way in the project 24GRD10 SmartGasNet.

Acknowledgment

This project has received funding from the Ministry of Economic Affairs of the Netherlands.

Statements

The data that support the findings of this study are available from the corresponding author upon request. The data are provided exclusively for non-commercial research purposes.

Author contributions
Conceptualization: FG, AvdV; Methodology: FG; Software: FG; Validation: FG; Formal analysis: FG; Investigation: FG; Resources: AvdV, EdB; Writing- Original draft: FG, AvdV, EdB; Visualization: FG, AvdV; Supervision: AvdV; Project administration: AvdV; Funding acquisition: AvdV

References

1. ISO 6145-10 Gas analysis – Preparation of calibration gas mixtures using dynamic volumetric methods – Part 10: Permeation methods (2002).
2. ISO 6145-7 Gas analysis – Preparation of calibration gas mixtures using dynamic volumetric methods – Part 7: Thermal mass-flow controllers (2018).
3. E. Flores, J. Viallon, P. Moussay, F. Idrees, and R. I. Wielgosz. Highly accurate nitrogen dioxide (NO2) in nitrogen standards based on permeation. Anal. Chem., **84**(23):10283 (2012).
4. H. Meuzelaar, J. Liu, S. Persijn, J. van Wijk, and A. M. H. van der Veen. Trace level analysis of reactive ISO 14687 impurities in hydrogen fuel using laser-based spectroscopic detection methods. Int. J. Hydrogen Energy, **45**(58):34024-34036 (2020).
5. E. Flores, et al.. International comparison CCQM-K74.2018: Nitrogen dioxide, 10 µmol/mol. Metrologia, **58**(1A):08018 (2021).
6. J. Viallon, E. Flores, F. Idrees, P. Moussay, R. I. Wielgosz, D. Kim, Y. D. Kim, S. Lee, S. Persijn, L. A. Konopelko, Y. A. Kustikov, A. V. Malginov, I. K. Chubchenko, A. Y. Klimov, O. V. Efremova, Z. Zhou, A. Possolo, T. Shimosaka, P. Brewer, and T. Macé. CCQM-K90, Formaldehyde in nitrogen, 2µmol mol^{-1} Final report. Metrologia, **54**(1A):08029-08029 (2017).
7. ISO 14687 Hydrogen fuel quality – Product specification (2025).
8. C. Chatfield and H. Xing. The analysis of time series. Chapman & Hall / CRC Texts in Statistical Science. CRC Press, London, England, 7th edition (2019).
9. R.H. Shumway and D.S. Stoffer. Time series analysis and its applications: with R examples. EDP Springer International Publishing, (2017).
10. A. M. H. van der Veen, F. Gugole, K. Folgerø, A. M. Skålvik, J. Kutin, G. Bobovnik, K. Rasmussen, L. C. Nordhjort Mjølna, and E. Venslovas. Best practices in the evaluation of the measurement uncertainty of quantities relevant to fiscal measurements along the hydrogen supply chain. (2025).
11. A. M. H. van der Veen, F. Gugole, K. Folgerø, A. M. Skålvik, J. Kutin, G. Bobovnik, K. Rasmussen, L. C. Nordhjort Mjølna, and E. Venslovas. Metering uncertainty for custody transfer of hydrogen for transport, heat, and storage. (2025).

Molecular Line Parameters, Impact on Climate Models and the Quantum-Pascal

Tom Rubin*, *André* Kussicke, *Sara* Gehauf, and *Olav* Werhahn,

Physikalisch-Technische Bundesanstalt, Braunschweig and Berlin, Germany

Abstract: Infrared absorption spectroscopy is one of the most powerful techniques for determining gas compositions. A key input parameter for this method is the absorption line strength, which also has a significant impact on atmospheric and climate models. Line strengths can either be determined experimentally, as demonstrated in this work for three of the strongest absorption lines of carbon monoxide (CO), or calculated using ab-initio theoretical methods. Improving and experimentally validating theoretical line strengths offers the fundamental advantage that complete absorption spectra can be calculated with higher accuracy, leading to more reliable spectroscopic knowledge for climate models. At the Physikalisch-Technische Bundesanstalt (PTB), CO absorption line strengths in the fundamental band in the 4.5 µm wavelength range (R8-R10) were experimentally determined at 296 K, yielding values of $4.452(16)\cdot 10^{-19}$, $4.217(16)\cdot 10^{-19}$, and $3.851(15)\cdot 10^{-19}$ cm molecule^{-1} respectively. These results show excellent agreement with the latest theoretical values recently reported by University College London, at sub-percent level. In contrast, line strengths currently used in widely applied databases such as HITRAN deviate from the experimental results by 0.66 %, 1.32 %, and 1.21 %, respectively.

1 Introduction

In this contribution we will exemplify recent achievements made during the Quantum-Pascal project in high-resolution tunable diode laser spectroscopy. To this end, we discuss new absorption line strengths measurements, performed at PTB. Available first data are for line strengths of carbon monoxide in its fundamental band in the 4.5 µm wavelength range. Our data is in excellent agreement with the values predicted by theorists from the University College London (UCL). This data serves as a benchmark for quantum chemistry calculations, which are, in turn, of fundamental relevance to current climate models. With the present measurement design, realized at the PTB, uncertainties of molecular line strengths are foreseen to achieve the sub permille level. Finally, we discuss the close links between the so-called Quantum-Pascal, gas metrology and climate action.

2 Theory

The Lambert–Beer law describes the attenuation of an electromagnetic wave propagating through an absorbing medium. The fractional loss in intensity, $\frac{I}{I_0}$, over an infinitesimal path length, dL, is given by:

$$\mathrm{d}I(\nu) = -\alpha(\nu)\, I(\nu)\, \mathrm{d}L \qquad (1)$$

where $\alpha(\nu)$ is the absorption coefficient given at the frequency ν. Neglecting scattering and other extinction processes, integration yields:

$$I(\nu) = I_0(\nu)\, e^{(-\alpha(\nu)\, L)} \qquad (2)$$

The absorption coefficient is proportional to the number density ρ. For a single absorption line centered at ν_0, this leads to:

$$I(\nu) = I_0(\nu)\, e^{(-\alpha(\nu-\nu_0)\, \rho\, L)} \qquad (3)$$

In tunable laser absorption spectroscopy (TLAS), the laser frequency is ideally scanned across an isolated absorption line, and the integrated absorbance is measured as:

$$A_{\mathrm{line}} = \int -\ln\left(\frac{I(\nu)}{I_0(\nu)}\right) \mathrm{d}\nu \qquad (4)$$

The frequency-integrated absorption coefficient defines the line strength parameter S as follows:

* Corresponding author: tom.rubin@ptb.de

$$S = \int \alpha(\nu - \nu_0)\, d\nu \qquad (5)$$

Combining equations (4) and (5) yields the integrated Lambert–Beer law in its logarithmic form.

$$A_{\text{line}} = S \rho L \qquad (6)$$

Using the ideal gas law, which is a sufficiently good approximation for the low pressures of several pascals used in this work:

$$\rho = P / (k_B T) \qquad (7)$$

with the gas temperature T, the Boltzmann constant k_B and the gas pressure P, the final experimental equation becomes:

$$A_{\text{line}} = S P L / (k_B T) \qquad (8)$$

Theoretical line strength parameters are temperature-dependent and are usually given for a reference temperature of 296 K. This temperature was used for the experimental assessment of S in this work, so no further consideration of temperature dependence is necessary.

3 Methods

A tunable laser absorption spectroscopy setup was optimized to measure several of the strongest absorption line strengths of CO. This setup was previously used to measure water vapour absorption line strengths with sub-percentage uncertainties [1]. It now utilizes a quantum cascade laser and a shorter single path cell (l=29 mm) which can be evacuated for reference measurements or filled with the absorbing CO gas. The emission frequency of the QCL is tunable over several CO absorption lines of the fundamental band between 2175 cm^{-1} and 2185 cm^{-1}. The frequency-dependent transmission signals measured by fast and highly sensitive photo detectors, are then converted into absorbance and finally numerically integrated to get the measurement parameter A_{line} (Eq. 8). The CO gas pressure and temperature are monitored using a calibrated Capacitive Diaphragm Gauge and multiple calibrated Standard Platinum Resistance Thermometers, respectively. This is summarized as 'Gas handling' in Figure 3.1. For the numerical integration of the absorption signal precise knowledge of the relative laser frequency tuning is required. Therefore, the QCL is frequency-locked to a high finesse, indirectly tunable Fabry-Perot resonator, (Scanning FPI in Figure 1) and additionally sent through an evacuated Herriott-Cell (TAC in Figure 1) with an optical path length of 6.4 m, which serves as an optical interferometer which serves as an etalon. Analyzing the fringe pattern of the interference signal gives rise to the absolute frequency tuning of the QCL during TLAS measurements. The experimental setup is integrated into a thermal enclosure stabilized to 296 K using an ultra-stable water bath in combination with several radiators and air fans.

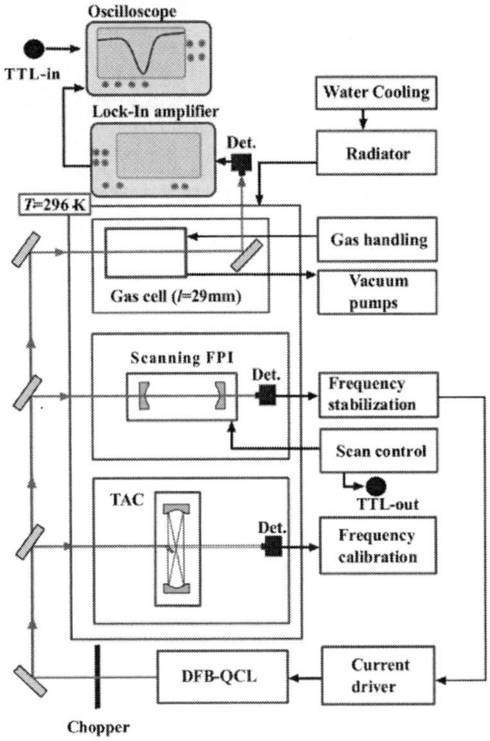

Figure 1: Experimental setup used to perform the line strength measurements for the three different CO absorption lines of the fundamental band.

4 Results

The absorbances of the R8, R9 and R10 absorption lines of the R-branch in the fundamental CO band were measured at different CO pressures between 5 Pa and 50 Pa. Figure 2 illustrates the R branch of the three CO isotopologues with the highest abundance ($^{12}C^{16}O$, $^{13}C^{16}O$ and $^{12}C^{18}O$) around 2150 cm^{-1}, where the absorption measurements where performed (data taken from the HITRAN API [2]).

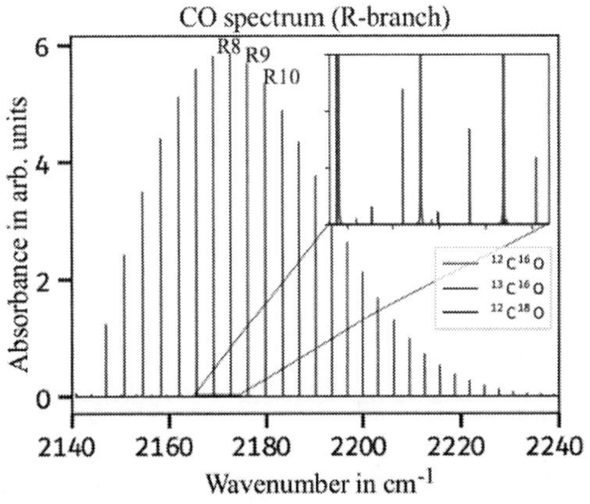

Figure 2: R branch of the three CO isotopologues between 2140 cm^{-1} and 2240 cm^{-1}. Using the TLAS setup in Figure 1, the line strengths of the R8, R9 and R10 absorption lines of the main isotopologue were investigated.

The experimentally obtained, pressure dependent line areas A_{line} (Eq. 8) were analyzed against $PL(k_BT)^{-1}$ as seen in Figure 3. The respective line strengths of the three absorption lines can be obtained from the slopes of the linear fits.

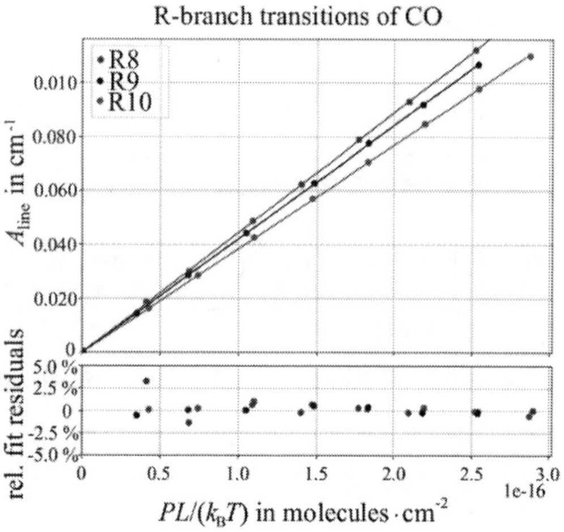

Figure 3: Pressure dependent CO line areas of the R8, R9 and R10 absorption lines. The residuals of the linear fits are shown below the experimental data.

The experimentally obtained line strengths values $\boldsymbol{S}$ are given in Table 1 together with their line positions and the expanded relative measurement uncertainties $\boldsymbol{\mu_{rel}(S)}$.

Table 1: Experimental results for S at 296.00(2) K

Line	Wavelength in cm^{-1}	S in 10^{-19} cm/molec	$\mu_{rel}(S)$ in % (k=2)
R8	2176.2835	4.452(16)	0.34
R9	2179.7719	4.217(16)	0.37
R10	2183.2238	3.851(15)	0.37

5 Discussion

The experimental results are compared with the theoretical values from the HITRAN2024 database [2] and with the latest calculations from the University College London (UCL) [3] in Table 2.

Table 2: S at 296 K in comparison with theoretical values from HITRAN [2] and UCL [3].

Line	S/S_{HITRAN} -1	S/S_{UCL} -1
R8	0.66 %	-0.35 %
R9	1.32 %	0.28 %
R10	1.21 %	0.14 %

The experimentally determined line strengths of the three absorption lines agree with theoretical calculations within the expanded uncertainty and the relative difference between experiment and calculation is better than 0.35 %. The difference between the experimental results presented in this paper and the respective line strengths listed in the HITRAN database is between 0.66 % and 1.32 %, suggesting that the line strengths in the database are slightly too large.

Table 3 shows the exemplary expanded uncertainty expanded budget for the R8 line strengths measurement. The highest contribution to the measurement uncertainty comes from the statistical noise of the data when performing the linear fit (Figure 3). Other significant factors that contribute to the overall measurement uncertainty are the gas pressure and temperature measurements, the gas purity of the commercial CO gas, the laser linewidths of the QCL, the measurement of the optical path lengths of the CO gas cell and finally the isotopologue mixture of the commercial CO gas.

Table 3: Uncertainty budget for R8 line measurement

Quantity	$\mu_{rel}(S)$ in % (k=2)	Type
Linear fit (S)	0.284	A
Pressure	0.120	B
Frequency calib.	0.126	B
Gas purity	0.030	B
Temperature	0.008	A
QCL linewidth	0.006	A
Optical pathlength	0.004	B
$^{13}C^{16}O$ contribution	0.004	B

The experiments reported in this paper are embedded in a larger framework at PTB. With the data's relevance for climate sciences, it belongs to the Innovation Cluster for Environment & Climate metrology [4]. With the application of spectroscopy, it also belongs to the QuantumPascal and MQB-Pascal projects [5, 6]. Detailing both aspects is beyond the scope of this paper. However, the determination of metrological traceable molecular line strengths is a core matter of metrology for climate sciences at PTB with various projects and other studies published before [7, 8, 9]. Line strength data impact climate sciences and related climate models wherever spectra and atmospheric transportation of radiation, i.e. sunlight matter. Networks of terrestrial, airborne, and satellite-borne measurement systems monitor the planet to determine whether climate change is driven by human activity or natural processes. All of these rely on the knowledge of molecular line data, and metrological traceability of an increasing number of them will guarantee the reliability and comparability of the inferred atmospheric conditions.

6 Conclusion

With CO being (only) a greenhouse gas precursor this molecular species often serves as a benchmark for quantum chemical calculations and new measurement approaches [10]. With only two atoms the line strengths are easier to calculate than those of more complex molecular structures directly relevant to climate sciences. Also, for new measurements, the comparably easy to handle CO in spectroscopic experiments provides valuable insight often improving the measurement process.

Improved uncertainties of CO absorption line strengths do come with direct relevance for ongoing metrology research projects. This includes, for example, the current joint research projects PriSpecTemp (22IEM03) [7] and MQB-Pascal (22IEM04) [6] in the European Partnership on Metrology programme [European Association of National Metrology Institutes (EURAMET e.V.), European Partnership on Metrology, [11]]. Within PriSpecTemp, ratios of multiple CO absorption lines are used to determine gas temperatures with high accuracy. This is achieved by means of remote sensing techniques. The reliability of such spectroscopic thermometry critically depends on accurate and metrological traceable line strength data.

In the context of the new realization of the pascal based on density measurements, the lowest uncertainties are currently achieved using Fabry–Perot refractometry (FPR) with pure gases such as helium, argon, and nitrogen. This approach relies on precise measurements of gas refractivity. In comparison, absorption by CO, as investigated in this work, produces a signal which is orders of magnitude stronger and therefore particularly well suited for measurements at very low pressures.

At present, FPR addresses a pressure range from approximately 1 Pa to 300 kPa. The lower limit of this range, around 1 Pa, corresponds to optimal signal-to-noise ratios in the tunable laser absorption spectroscopy (TLAS) measurements presented here for an optical path length of about 1 m. Using the same experimental setup, optical path lengths of up to 100 m have been realized, enabling optimal measurements in the pressure range around 10 mPa.

By employing cavity-based techniques such as cavity ring-down spectroscopy (CRDS) and cavity-enhanced absorption spectroscopy (CEAS), effective optical path lengths of several kilometers can be achieved. This allows precise measurements of correspondingly lower CO partial pressures, extending well into the ultra-high vacuum regime (pressures below 10^{-6} Pa). As a result, improved uncertainties established by FPR at higher pressures can, at least partially, be transferred to significantly lower pressure ranges.

Furthermore, our line strength results will help improve future CO measurements related to climate sciences and environmentally relevant measurements, e.g., in traffic exhaust assessments.

Acknowledgements

This work was part of the 22IEM04 MQB-Pascal project, which has received funding from the European Partnership on Metrology, co-financed from the European Union's Horizon Europe Research and Innovation Programme and by the Participating States.

References

1. T. M. Rubin, M. Sarrazin, N.F. Zobov, J. Tennyson, and O. L. Polyansky, Sub-percent accuracy for the intensity of a near-infrared water line at 10,670 cm− 1: experiment and analysis. *Molecular Physics*, **120**(19-20), e2063769, (2022), https://doi.org/10.1080/00268976.2022.2063769
2. I. E. Gordon, L. S. Rothman, R. J. Hargreaves, F. M. Gomez, T. Bertin, C. Hill, et al.,"The HITRAN2024 molecular spectroscopic database", *J. Quant. Spectrosc. Radiat. Transfer*, in press, (2026), doi:10.1016/j.jqsrt.2026.109807.
3. K. Bielska, A. A. Kyuberis, Z. D. Reed, G. Li, A. Cygan, R. Ciuryło, E. M. Adkins, L. Lodi, N. F. Zobov et al., Subpromille Measurements and Calculations of CO (3–0) Overtone Line Intensities, *Phys. Rev. Lett.* **129**, 043002 – Published 21 July, (2022), https://doi.org/10.1103/PhysRevLett.129.043002
4. https://www.ptb.de/environment-climate
5. https://www.ptb.de/empir2019/quantumpascal/
6. https://www.mqb-pascal.ptb.de/home
7. https://www.prispectemp.ptb.de/home
8. https://www.eumetrispec.org/
9. https://www.metctg.ptb.de/home
10. J. T. Hodges, K. Bielska, M. Birk, et al., International Comparison CCQM-P229: Pilot Study to Measure Line Intensities of Selected 12C16O Transitions, https://www.bipm.org/documents/d/guest/ccqm-p229
11. https://www.metpart.eu/

https://doi.org/50.50(5/epjconf/10163(10100(

Bosch Optical Gas Spectrometer - A New Dimension in Gas Measurement with Raman Spectroscopy

Alexander Stratmann*

Robert Bosch GmbH, PS-SB/EPN, Wernerstr. 51, 70469 Stuttgart, Germany

Abstract. Analysing the composition of a gas is a complex challenge. Depending on the use case and considering thermodynamic constraints such as gas temperature, gas pressure, and flow rate, various technologies are available. Specific sensors are usually straightforward to use. However, they often have limited concentration ranges and require more or less complex calibration. More sophisticated techniques, such as mass and absorption spectroscopy or gas chromatography, each have their own limitations. Some require ambient pressure, dedicated probe extraction, or time-consuming analysis processes. Raman spectroscopy appears to overcome many of these limitations. However, the Raman effect is so weak that complex and expensive setups are required. To overcome these problems, Bosch developed the BOGS (Bosch Optical Gas Spectrometer). Raman gas spectroscopy is now extremely easy. Gas concentrations as low as 100 ppm can now be detected, and even lower concentrations when the pressure is higher (up to 40 bar) and the detection time is increased from seconds to minutes.

1 Raman Spectroscopy

Raman spectroscopy, an inelastic light scattering, is an effective method for measuring all molecular gases. This means that all species besides noble gases have a specific Raman signal. As stated in the publication by the discoverers [1], the Raman effect is very weak, especially for gases due to their low density. This is one of the main reasons why Raman spectroscopy is not often used to analyse gases in industry. It is mainly because a complex and expensive laboratory setup is required. The challenge in introducing Raman spectroscopy as a measurement device lies in overcoming the sensitivity issue and making the setup more industrialised. The BOGS (Bosch Optical Gas Spectrometer) is a significant step towards achieving that goal.

1.1 Technical Setup

We have chosen the classic 90-degree setup, as shown in Figure 1. The Raman excitation is done by a high-power laser diode with constant approx. 3.5 W at 450 nm. The scattered light passes through a long-pass filter to reduce disturbing Rayleigh light, before being detected by a Czerny-Turner spectrometer with a 512-pixel CCD detector. A Raman shift from approximately 395 to 5050 cm^{-1} can thus be detected. The spectral range above 3500 cm^{-1} is not often available and enables measurements of molecules such as water vapour at approximately 3650 cm^{-1} and hydrogen at approximately 4150 cm^{-1}.

The gas to be measured passes through a stainless-steel pipe containing sapphire windows, through which the laser beam is guided, and the scattered light is collected. As with Raman spectroscopy in general, no carrier gas is required. Nor is gas flow necessary.

1.2 Bosch Optical Gas Spectrometer

The BOGS (Figure 1) is a newly developed gas analyser based on Raman spectroscopy. We have industrialised this technology and integrated it into a compact tabletop device, with a 19-inch housing.

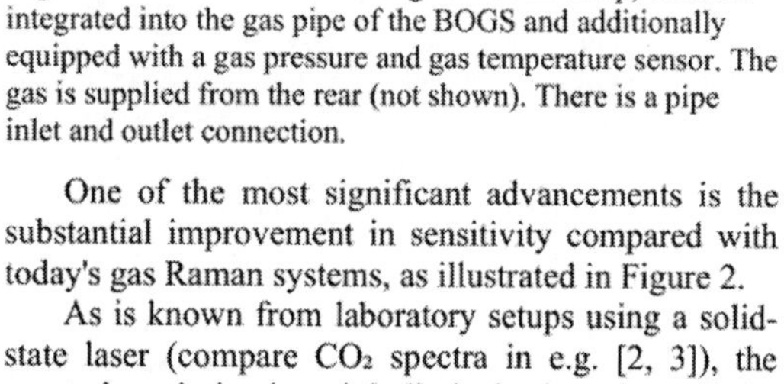

Fig. 1. Sketch of a classic 90-degree Raman setup, which is integrated into the gas pipe of the BOGS and additionally equipped with a gas pressure and gas temperature sensor. The gas is supplied from the rear (not shown). There is a pipe inlet and outlet connection.

One of the most significant advancements is the substantial improvement in sensitivity compared with today's gas Raman systems, as illustrated in Figure 2.

As is known from laboratory setups using a solid-state laser (compare CO_2 spectra in e.g. [2, 3]), the spectral resolution is mainly limited using a laser diode. Nevertheless, this is sufficient for a lot of applications.

* Corresponding author: alexander.stratmann@de.bosch.com

https://doi.org/50.50(5/epjconf/10163(10100(

In summary, depending on the test conditions and anticipating the next chapter, the BOGS can measure concentrations ranging from 100 ppm to 100 vol%, and it can operate at gas temperatures of up to 40 °C and pressures of up to 40 bar, as well as sub-ambient pressures. Thanks to its highly sensitive Raman signal, even evaporated liquids can be studied.

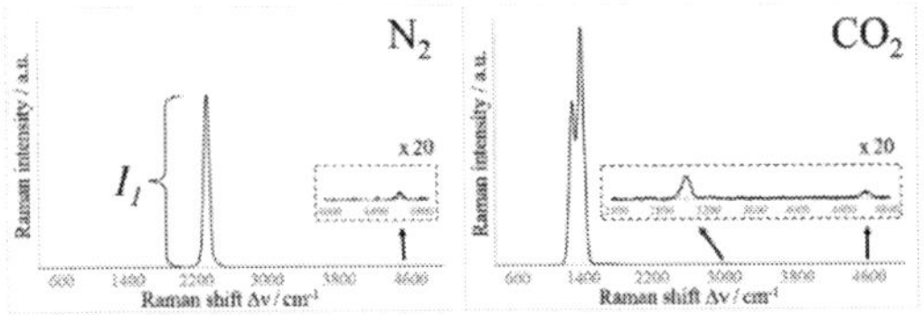

Fig. 2. With BOGS measured Raman spectra of pure nitrogen and carbon dioxide, shown as the median of three 5-second measurements at 3500 hPa. At 20x magnification, the second-order Raman lines can be clearly seen. I_I is here an evaluation example of the peak height as an indicator of the nitrogen intensity.

1.3 Gas Concentration & Calibration

The radiation flux Φ_I of one Raman signal (at wave-number ν_I) is strong linear to the gas density, respect-tively to the gas particles N_I in the observed volume V [2, 3]:

$$\Phi_I(\nu_I) = N_I / V \cdot \partial\sigma(\nu_I) / \partial\Omega \cdot \Omega \cdot l \cdot P_L = a_I \cdot N_I / V. \quad (1)$$

Table 1. Formula symbols at section 1.3 that have not been described in the text

$\partial\sigma(\nu)/\partial\Omega$	differential cross-section
Ω	solid angle of observation
l	length of scattering volume
P_L	laser power
a, b	constants
α	Raman sensitivity factor/constant
k	Boltzmann constant

This enables the evaluated intensity I_I (see Figure 2) to be calibrated at a single point, depending on the exposure respectively detection time t of the measure-ment:

$$I_I = b_I \cdot \Phi_I(\nu_I) \cdot t = b_I \cdot a_I \cdot N_I / V \cdot t = \alpha_I \cdot N_I / V \cdot t. \quad (2)$$

The concentration c_I (in vol%) according to ISO 14912 with $c_i \stackrel{\text{def}}{=} V_i / V$ and $c = \Sigma_i V_i = 1$ of a gas species can be calculated by measuring the gas pressure p, the gas temperature T and applying the ideal gas law

$$p \cdot V = N \cdot k \cdot T \Rightarrow N = p \cdot V / (k \cdot T) \quad (3)$$

finally by

$$c_I = I_I / (\alpha_I \cdot t) \cdot k \cdot T / p. \quad (4)$$

Therefore, to calculate the gas concentration, the Raman sensitivity or calibration factor α of the Raman peak of each species must be known. There are many ways to determine or calibrate this factor. Some of these methods are presented in the following section.

2 Measurements & Use Cases

The in following shown measurements, taken using the Raman spectroscopic system implemented in the BOGS, typically are normalised spectra at a standard pressure of 1013.25 hPa, a temperature of 20 °C and an exposure time of 1s. They are calculated using equation (3), which uses the ratio of the measured gas pressure and temperature to the standard and the exposure time ratio to 1 s. The advantage here is that it allows the same evaluation algorithm to be used in different conditions relating to gas pressure, temperature, and exposure time. It also makes it easier to compare the important signal-to-noise ratio. A ratio of more than 3:1 is usually required for a significant detection threshold.

2.1 Fundamentals

Figure 3 shows the normalised spectra obtained by measuring dry and ambient air.

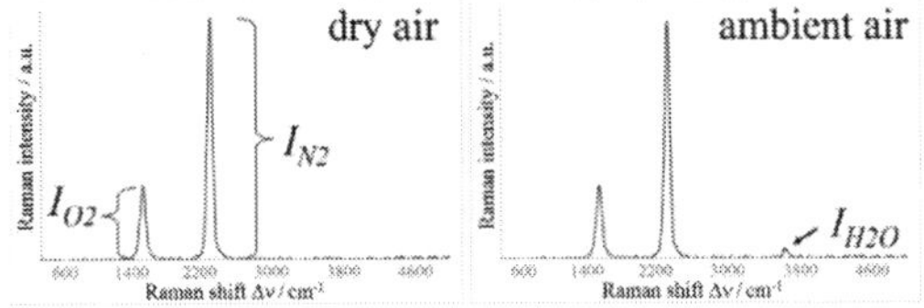

Fig. 3. Raman spectra of dry air and ambient air (including moisture) measured over 10 s at 1150 hPa, 32 °C and 0.4 l/min gas flow.

At altitudes of up to around 80 km, the composition of dried ambient air is constant to 78.09 vol% nitrogen, 20.95 vol% oxygen and 0.96 vol% other gases (e.g. argon and carbon dioxide) [4]. This data can be used to calibrate α_{N_2} and α_{O_2} from equation (4), with the intensity I_{N2} and I_{O2} being calculated either from the peak height or, for better repeatability, from the peak area of the single, positioned Raman peaks. Further-more, measuring ambient air afterwards gives the nitrogen and oxygen concentrations, which has been reduced by moisture. The same ratio was used to reduce the concentrations of the other gases. The final difference to 100 vol% is than the water vapour concentration, which is 1.85 (abs.) vol% here. Therewith the α_{H_2O} can be calculated by the water Raman peak intensity evaluation (German patent application DE 10 2023 211 698 A1).

The BOGS software includes these and more options for Raman signal evaluation and gas concentration calibration.

One of most important topic is the accuracy of a measurement. Depending on the definition used for different use cases, we can provide data on repeatability

here. Each gas species has different sensitivities (e.g. methane is 10 times more sensitive than nitrogen), a longer detection time (limited by the CCD detector dynamic), and a higher pressure will increase the signal-to-noise ratio. Overlapping Raman lines require more complex evaluation algorithms. This means that no single repeatability value can be defined. Nevertheless, we will provide two examples: An absolute and a relative gas concentration evaluation.

A repetition measurement of a single gas (here hydrogen), also comparable of a single component in an unknown mixture, gives a standard deviation of the hydrogen concentration of $\sigma_{n-1} = 0.12$ vol%. This was calculated by evaluating the peak area of the vibrational Raman line from a median average of five measurements taken every 15 minutes over 60 repetitions: see Figure 4 (left).

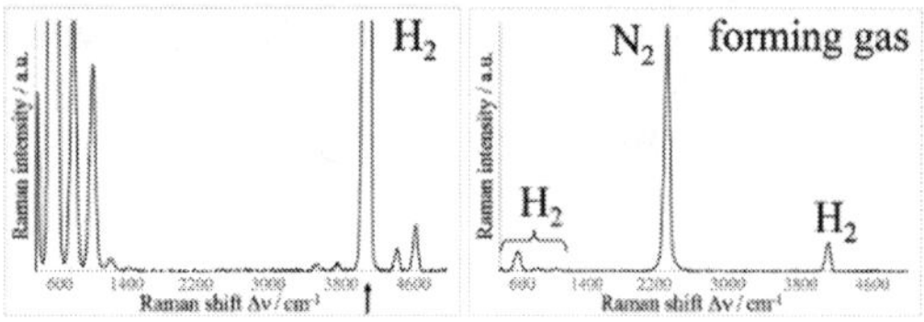

Fig. 4. Zoomed-in view of the Raman spectrum of hydrogen 5.0 (99.999) and 4 vol% nitrogen in hydrogen 5.0 (≙ forming gas) measured over 5 s at hydrogen, 10 s at forming gas, and 1150 hPa, 32 °C, 0.4 l/min for both. The peak marked on the left is the vibrational; all the others are the rotational hydrogen Raman lines.

If the gas matrix is known, as with forming gas (see Figure 4 on the right), repeatability increases significantly. This means that the total concentration of all known components can be normalised to 100 vol%. Minor influences, such as pressure and temperature measurement deviations or gas flow disturbances, will be eliminated. The results in Table 2 were obtained by initially calibrating the peak area with the forming gas, followed by 40 repetitions at 15-minute intervals. The above-described evaluation also normalises the gas concentration to 100 vol%.

Table 2. Result of a relative/normalised gas concentration evaluation of a repetition measurement taken 40 times at nearby ambient air pressure.

gas species	gas concentration / vol%	measured mean value / vol%	standard deviation / vol%
N_2	96.00	95.98	0.03
H_2	4.00	3.99	0.02

In summary, an absolute measurement results in a standard deviation of around 0.1 vol%, whereas a relative evaluation results in a value one order of magnitude lower, at around 0.01 vol%. That is at ambient pressure. Increasing the gas pressure will improve repeatability even further. We are not yet able to provide any concrete results, but we will do so as soon as we are able. Increasing the exposure time is here not possible. Due to the limited CCD detector dynamic, we have chosen the 5 and 10 s as the maximum possible Raman signal height.

2.2 Hydrogen & ISO 14687

In [5], a good overview of the measurement possibilities for impurities, in accordance with ISO 14687, is provided. No Raman technology was used at that time. Figure 5 shows our initial approaches with BOGS.

At approximately 5000 hPa, the two most significant impurities can be detected with a required signal-to-noise ratio of 3:1. Please note that the ISO limit value of 100 ppm methane would already be reached at ambient pressure.

And, at a typical electrolyser working pressure, oxygen and water can also be observed in situ. The minimum detectable concentration, which is also limited by the shown Raman peak overlap, is now under development.

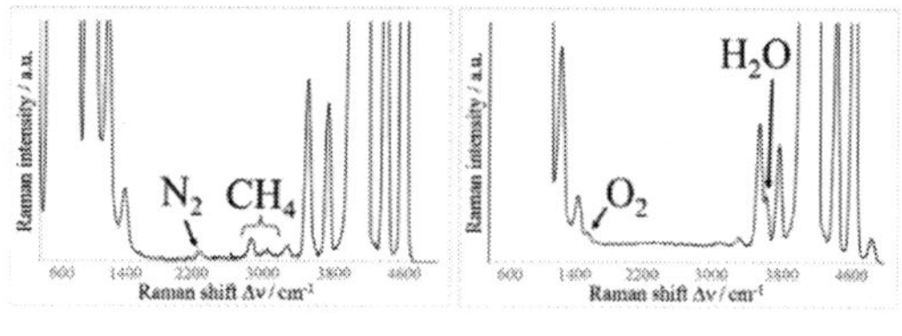

Fig. 5. Zoomed-in view of the median average of five hydrogen spectra, with 288 ppm of nitrogen and 98 ppm of methane impurities, is shown on the left, measured over 60 seconds at 5000 hPa and 32 °C. On the right is the hydrogen measured at an electrolyser with approximately 600 ppm of oxygen and 200 ppm of water vapour impurities, measured over 30 seconds at 4 MPa and 32 °C.

Finally, two questions remain. Firstly, is it possible to measure the other constituents according to ISO 14687, and if so, at what detection limit? For that purpose, we started a research project in cooperation with the institute Laseranwendungstechnik at Ruhr-University Bochum and Linde Gase at Munich with the public funding project HyQ2Ra. This is sponsored by the German Federal Ministry for Economic Affairs and Energy. The project has ended in December 2025. Preliminary results are shown in [6, 7]. Anticipating the report: Most of the constituents have Raman lines in the hydrogen-free spectral area (see Figure 4 on the left: 1700 - 3500 cm^{-1}), making them easier to evaluate. Increasing the gas pressure above the current 4 MPa achieved with BOGS also helps to increase sensitivity. The known Raman line broadening and shifting seem to be negligible up to 10 MPa, or even higher.

2.3 Natural Gas & THC

The knowledge of the gas compensation of natural gas is essential (billing, gas motor control, …). Today, gas chromatography is usually considered to be state of the art. The advantages of Raman technology include the ability to take measurements at higher pressures (for example, at the transport grid, pressures can reach up to 10.5 MPa) without having to extract the probe and the speed of the process (less than a few minutes). A first approach with a classic Raman spectroscopic setup is given in [8].

We have developed an optimised evaluation algorithm using the BOGS by fitting the previously measured spectra of pure gas substances to those of a natural gas mixture. Figure 6 shows the spectra of the alkanes that are included in the evaluation. Isomers have different spectra and can therefore be distinguished. By the way, isotopes, such as those from hydrogen, are also different. Additional, for the natural gas evaluation, nitrogen, oxygen, carbon dioxide and hydrogen are included.

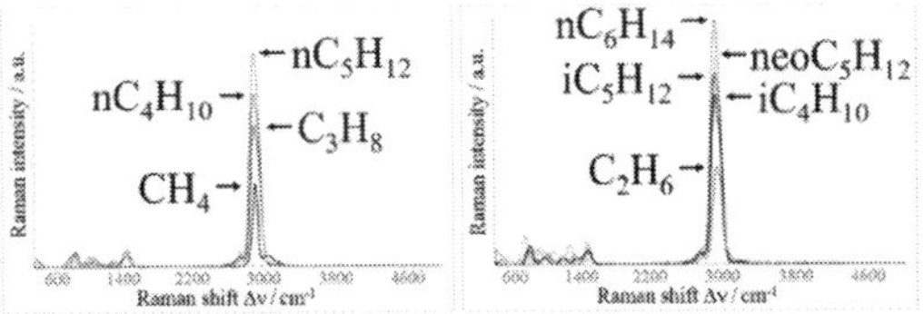

Fig. 6. Raman spectra of alkanes and their isomers from methane (CH_4) to n-hexane (nC_6H_{14}), scaled to the same gas concentration. The higher the atoms in the molecule, the higher the Raman signal/sensitivity.

To verify the algorithm and ensure consistency with the OIML R 140 and ISO 6967 standard, special calibration gases must be tested. Figure 7 shows a typical measurement of a calibration gas called '9M'.

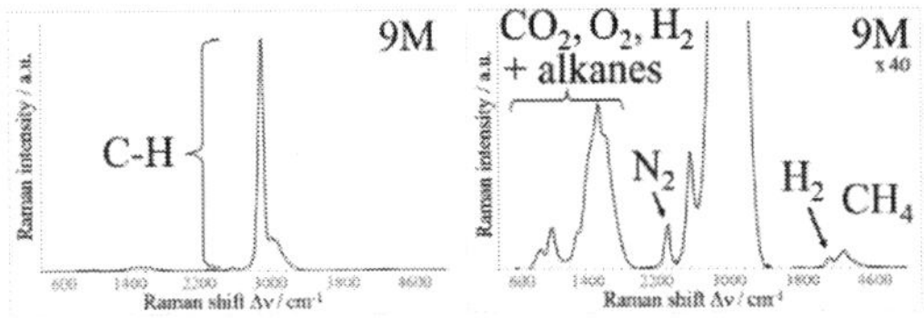

Fig. 7. Raman spectrum of the 9M calibration gas for natural gas. The (normalised) spectrum was created using higher detection times outside the dominant Raman C-H vibration area (2400 - 3500 cm^{-1}) to improve the signal-to-noise ratio. The zoom-in view on the right shows the areas of interest outside the C-H vibration. The median average of five measurements taken over a period of 21 seconds at a pressure of 3500 hPa and a temperature of 26 °C is shown.

The results of the normalised concentration evaluation, as described in the 'Fundamentals' section, are shown in Table 3. The difference between the measurement and the certification is very small. With a calorific value deviation of -0.43 %, it is below the ±0.5% limit. This demonstrates the accuracy of the measurement, and repeating the measurement (not shown) produces the same result.

Further calibration gases are currently under investigation. However, the initial results demonstrate the significant potential of Raman spectroscopy, even when dealing with more complex spectra featuring overlapping peaks and numerous gas components.

Figure 6 shows the Raman spectra of the alkanes. There is one more thing we can learn from these spectra. All hydrocarbons have a very sensitive Raman signal in the 3000 cm^{-1} area; this is due to the C-H vibration of the molecules. If a significant signal is detected here (with a signal-to-noise ratio of more than 3:1), then hydrocarbons are present in the gas. This can be used as a detection threshold for determining the presence of THC (total hydrocarbons). It is, at a minimum, methane because the Raman signal for the hydrocarbon is the lowest.

Table 3. Gas concentration and measurement of the 9M calibration gas in vol%

gas species	9M gas certification	9M BOGS measurement
N_2	04.00	04.07
H_2	00.20	00.14
CO_2	02.48	02.37
O_2	00.40	00.57
CH_4	89.03	89.29
C_2H_6	02.49	02.30
C_3H_8	01.00	00.96
nC_4H_{10}	00.20	00.12
iC_4H_{10}	00.20	00.12
nC_5H_{12}	00.00	00.00
iC_5H_{12}	00.00	00.00
$neoC_5H_{12}$	00.00	00.00
nC_6H_{14}	00.00	00.06

2.4 Biogas

Lastly, as a further example of evaluation, we can consider a typical biogas Raman spectrum: Figure 8. The main components of this mixture are methane and carbon dioxide, along with small amounts of hydrogen and residual moisture. Although the Raman peaks of carbon dioxide and methane overlap slightly, the classic evaluation of peak area with the left half assigned to carbon dioxide and the right half to methane is sufficient for the required accuracy in the 0.5 vol% range. The small amount of hydrogen sulfide positioned directly on a side peak of methane is more complex. In addition to the above-described algorithm for natural gas, a simple fit of pure methane in the spectral area of 2600 - 3600 cm^{-1} can be used as a background to determine the residual hydrogen sulfide.

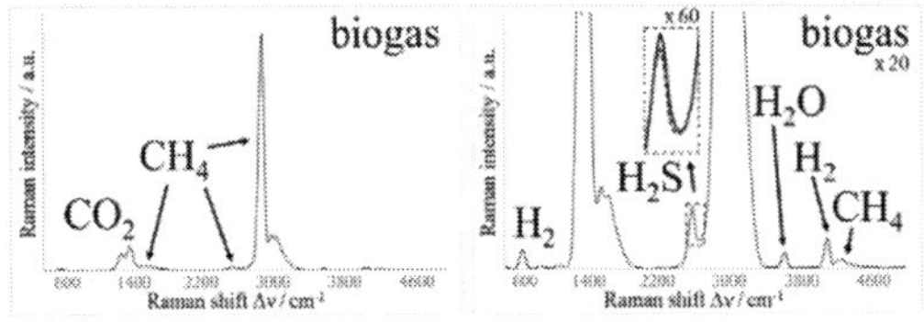

Fig. 8. Raman spectrum of biogas at ambient pressure. The two main components are 40 vol% carbon dioxide and 58 vol% methane, as shown on the left. In the zoomed-in view on the right, there is 1 vol% hydrogen and residual moisture. The 60x magnification shows the presence of 1000 ppm of hydrogen sulfide (black line) compared to its absence (dotted line).

One of the next steps in our investigation is to determine the detectable hydrogen sulfide concentration limit below 1000 ppm, which also depends on the gas pressure.

3 Summary & Outlook

The BOGS brings a new dimension to Raman spectroscopy for gases. We have demonstrated its various applications here. Many new applications, as well as previously impossible ones, are now achievable. While it does not overcome all limitations, it provides additional capabilities, particularly for molecules such as nitrogen, oxygen and hydrogen. In-situ and online applications at higher pressure can now also be easily implemented without any problems.

We have demonstrated classic and more sophisticated algorithms for quantitative spectral analysis. In addition to natural gas analysis, which involves separating overlapping Raman peaks to enable quantified measurements, more advanced evaluation algorithms are required. This is particularly important for applications involving low-concentration gas species.

According to ISO 14687, the public funding project HyQ2Ra evaluated increasing the pressure and using specific algorithms. In anticipation of the final report, it is thought that gas concentration in the one or two-digit ppm range can be achieved at higher pressures, potentially up to 10 MPa.

References

1. C. Raman, K.A. Krishnan, A New Type of Secondary Radiation, Nature **121**, 501–502 (1928).
2. A. Stratmann, G. Schweiger, Fluid Phase Equilibria of Ethanol and Carbon Dioxide Mixtures with Concentration Measurements by Raman Spectroscopy, Applied Spectroscopy **56 (6)**, 783-788 (2002). https://doi.org/10.1366/000370202760077531
3. A. Stratmann, G. Schweiger, Chapter 1.4 - Fluid Phase Equilibria of Binary Mixtures with Supercritical Solvents with in-situ Concentration Measurements by Raman Spectroscopy, 85-120 in G. Brunner, Supercritical Fluids as Solvents and Reaction Media (Elsevier, 2004). https://doi.org/10.1016/B978-044451574-2/50004-3
4. W. Roedel, T. Wagner, Physik unserer Umwelt: Die Atmosphäre (Springer-Verlag, Berlin, 2011). https://doi.org/10.1007/978-3-642-15729-5_1
5. L. Appold, A. Stratmann, Hydrogen, the Future in Mobility - Monitoring Gas Quality and Quantity by Using an Innovative Sensor, in Proceedings of the 23. Internationales Stuttgarter Symposium, Automobil und Motorentechnik, Forschungsinstitut für Kraftfahrwesen und Fahrzeugmotoren, FKFS, 344-348 (2023). https://doi.org/10.1007/978-3-658-42048-2_23
6. C.R. Günther, B. Haske, L. Appold, A. Stratmann, A. Ostendorf, C. Esen, Online hydrogen quality control for fuel stations using Raman spectroscopy, Optical Technology and Measurement for Industrial Applications Conference 2025; 137040P (2025) https://doi.org/10.1117/12.3075253
7. C. Beurey, B. Gozlan, M. Carré, T. Bacquart, A. Morris, N. Moore, K. Arrhenius, H. Meuzelaar, S. Persijn, A. Rojo A, A. Murugan, Review and Survey of Methods for Analysis of Impurities in Hydrogen for Fuel Cell Vehicles According to ISO 14687:2019, Front. Energy Res. 8:615149 (2021). https://doi.org/10.3389/fenrg.2020.615149
8. D.V. Petrov, I.I. Matrosov, A.R. Zaripov, A.S. Tanichev, Raman Natural Gas Analyzer: Effects of Composition on Measurement Precision, Precision, Sensors **22**, 3492 (2022). https://doi.org/10.3390/s22093492

EPJ Web of Conferences **352**, 0100(2101(6   )Ht sp:/ odor. :g0ig05g:et jconf:101(3510100(
GAS 2026

Accelerating Instrument Troubleshooting: An AI-Driven Approach to Eliminating O_2 Trap Breakthrough in GC-HDID Analysis

Chun-ling Chang [1*], *Hsien-Jen* TSENG[1], and *Syusheng* SIE[2]

[1]Air liquide, Quality Department, APAC EAP ALFE, Taiwan (R.O.C.)
[2]Air liquide, Production Department, APAC EAP ALFE, Taiwan (R.O.C.)

Abstract. The analysis of trace impurities in ultra-high purity oxygen (UPO) via GC-HDID is critically dependent on the O_2 trap's performance, where recurring breakthroughs pose a significant challenge. This paper presents an innovative troubleshooting framework that dramatically accelerates problem resolution by integrating AI, contrasting sharply with traditional, time-consuming methods. Conventional approaches involve laborious cycles of manual review, broad internet and literature searches, and multiple consultations before a testable solution can be proposed. Our novel workflow bypasses these inefficiencies. We fed instrument manuals and raw failure data from a 'T Instrument' directly into an AI model (Gemini) for a "deep research" phase. However, initial AI outputs were too broad, identifying general-purpose purifiers rather than the specific, regenerable O_2 trap in question, as many manufacturers do not disclose these proprietary details. Here, operator expertise became crucial. We iteratively tuned the AI's research, refining our queries to focus on the specific context of regenerable, copper-based catalysts used for GC matrix removal. This expert-guided "deep research" successfully filtered out irrelevant information and led the AI to confirm a universal, underlying chemical principle—the copper redox reaction ($2Cu+O2\rightarrow2CuO$; $CuO+H2\rightarrow Cu+H2O$)—across different manufacturers' traps. This pivotal, AI-generated insight, achieved through expert-led refinement, enabled a swift and accurate diagnosis when combined with operator experience. The root cause was not the regeneration reaction itself, but the incomplete removal of its H_2O byproduct. The solution was therefore clear: significantly extend the total duration of the high-temperature helium purge across both the 'Heater' and 'Standby' phases. This optimized protocol completely eliminated breakthrough events. This AI-augmented methodology, where human expertise directs AI's powerful analytical capabilities, represents a paradigm shift, saving considerable time on unfocused research and meetings, and presents a powerful, transferable framework for rapidly solving complex instrumentation challenges in gas analysis.

* Corresponding author: jenna.chang@airliquide.com

1 Introduction

In the semiconductor and electronics industries, the demand for ultra-high purity (UHP) gases is paramount. Gas Chromatography with a Helium Ionization Detector (GC-HDID) is a cornerstone technology for ensuring the quality of these gases, capable of detecting impurities at the parts-per-billion (ppb) level[1]. A critical component in the analysis of UHP oxygen (UPO) is the oxygen (O_2) trap, which removes the bulk oxygen matrix to prevent it from reaching and saturating the detector, thereby allowing for the accurate measurement of trace impurities like nitrogen (N_2), argon (Ar), and methane (CH_4)[2,3].

However, the O_2 trap is a consumable component with a finite lifespan. When it becomes saturated, a phenomenon known as "breakthrough" occurs, where oxygen is no longer effectively removed and begins to elute into the detector. This leads to a significant increase in baseline noise, quenching of the detector signal, and ultimately, the inability to produce reliable analytical results. Troubleshooting O_2 trap breakthrough has traditionally been a complex and time-consuming process. It often involves a lengthy, trial-and-error cycle of consulting instrument manuals, searching through vast online resources and scientific literature, and engaging in prolonged discussions with instrument manufacturers or senior technical staff. This conventional approach can lead to extended instrument downtime, impacting production schedules and quality control.

This paper presents a novel and accelerated approach to troubleshooting O_2 trap breakthrough in a GC-HDID system. By leveraging the capabilities of a large language model (LLM), Gemini, we developed a streamlined workflow that integrates AI-powered "deep research" with the indispensable expertise of human operators. This hybrid methodology was applied to a real-world case of O_2 trap failure in a 'T Instrument' used for UPO analysis. We demonstrate that this AI-driven approach can significantly reduce the time to resolution from over a week to less than 24 hours, offering a powerful new paradigm for instrument maintenance and troubleshooting in high-purity gas analysis.

2 Methodology

The core of our accelerated troubleshooting framework is a synergistic loop between AI-driven data analysis and iterative refinement by an experienced operator, designed to rapidly move from a general problem to a specific, actionable solution. The process is illustrated in Figure 1.

Fig. 1. AI-Driven Troubleshooting Flowchart

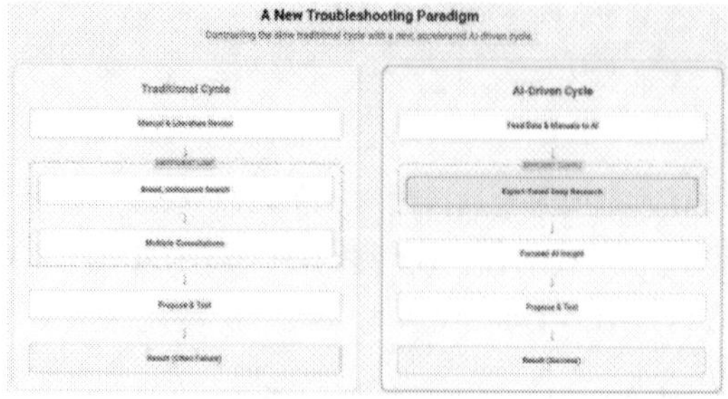

2.1 Initial Data Input and Problem Definition

The process began by providing the AI model (Gemini) with a comprehensive set of initial data. This included the complete operation and service manuals for the 'T Instrument' and raw chromatographic data logs showing clear evidence of O_2 trap failure.[4] The failure manifested as a progressive decline in the number of successful injections after each regeneration cycle, as shown in Figure 2. Data logs exhibited a rising baseline and suppressed peaks for key impurities (N_2, Ar), eventually leading to analytical failure..

Fig. 2. T-Instrument performance decline, showing the decreasing number of injections before O_2 trap failure

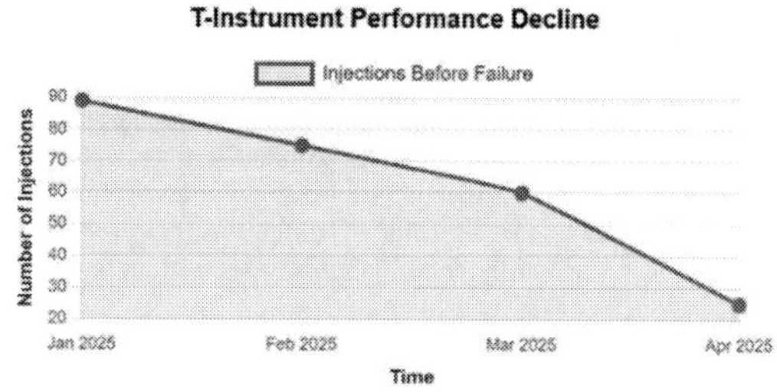

2.2 Hypothesis Formulation and AI-Assisted Verification

The initial AI outputs were too general, identifying purifier exhaustion but recommending the replacement of generic, non-regenerable purifiers.[2] This was due to the proprietary nature of the specific regenerable O_2 trap, for which manufacturers often do not disclose detailed chemical compositions.

At this stage, operator expertise became critical. We formulated a core hypothesis: despite different manufacturers (the T Instrument uses a VICI-type trap, while a successfully operating G Instrument uses an BTS column), the underlying chemical principle for regeneration was likely identical. We then tuned the AI's research with highly specific queries focusing on regenerable O_2 traps for GC matrix removal, particularly those using copper-based catalysts.[5,6] By cross-referencing information from supplier Material Safety Data Sheets (MSDS) with the AI's research, we confirmed the hypothesis. The AI deduced the universal mechanism to be a copper redox reaction:

- **Oxygen Removal:** 2Cu+O2→2CuO
- **Regeneration:** CuO+H2→Cu+H2O

This confirmed that the two trap types shared the same chemical basis, making their regeneration conditions directly comparable. The key to successful regeneration, therefore, was not only the completion of the reduction reaction but also the thorough removal of its water byproduct.

2.3 Comparative Analysis and Solution Synthesis

Based on the shared chemical principle, we performed a comparative analysis of the regeneration conditions between the successfully operating G Instrument and the failing T Instrument. The primary difference was not the regeneration gas (H_2) or temperature, but the duration of the subsequent high-temperature inert gas purge designed to remove the H_2O byproduct. The G Instrument employed a significantly longer heated helium purge phase.

This analysis pinpointed the root cause of the T Instrument's failure: an insufficiently long heated purge time, which left residual moisture on the copper catalyst. This moisture blocked the active sites, preventing effective oxygen removal and causing premature breakthrough.

The solution was to synthesize a new, optimized protocol for the T Instrument by drastically extending the total duration of the high-temperature helium purge across both the 'Heater' and 'Standby' phases. The final optimized parameters, derived from this AI-assisted comparative analysis, are detailed in Table 1.

Table 1. Optimized O_2 Trap Regeneration Conditions for the T Instrument

Parameter	Value
Regeneration Gas	Helium (He)
Flow Rate (mL/min)	90
Temperature (°C)	350
Duration (hr)	10
Standby Time (hr)	5
Total Time (hr)	15

3 Results and Discussion

The implementation of the AI-derived regeneration protocol yielded immediate and significant improvements in the performance of the T Instrument.

3.1 Restoration of Performance

The implementation of the AI-derived protocols yielded two distinct successful outcomes ("Outcome 1" and "Outcome 2"), which were validated against the benchmark performance of the Orthodyne system ("Reference"). The comparative results, detailed in **Table 2**, reveal that due to the hardware temperature limit of 210°C on the VICI traps, stability could only be achieved by compensating with **extreme purge duration** or **strict capacity management**.

Outcome 1: The "Long Cycle" (Instrument G) For Instrument G, the "Ghost Peak" issue caused by residual water was resolved not by temperature (which was capped at 190°C), but by time. By extending the heated purge phase from 12 hours to **21 hours** and the standby phase to 24 hours (Total Cycle ~48 hours), the baseline disturbances were completely eliminated. This confirms that at lower temperatures, water desorption is kinetically slow and requires a massive extension of the purge window.

Outcome 2: Capacity Management (Instrument T) For Instrument T, "Breakthrough" was managed by acknowledging the thermodynamic limits. Even with an extended purge (3-4 days), the trap at 190°C could not achieve the deep regeneration seen in the reference unit. The solution was "Discipline": establishing a fixed safety limit of **55 injections** per cycle to regenerate the trap before the shallow surface capacity was exhausted.

Reference Standard (Instrument O) The comparison with the Orthodyne unit (Inst. O) highlights the root mechanism. As shown in the "Reference" column of Table 2, the Orthodyne unit achieves stability in just 24 hours because it utilizes a regeneration temperature of **350°C**. This validates the "Time-for-Temperature" hypothesis: since the Agilent/VICI units cannot reach the kinetically efficient 350°C, they must substitute extreme purge durations to achieve the same baseline stability.

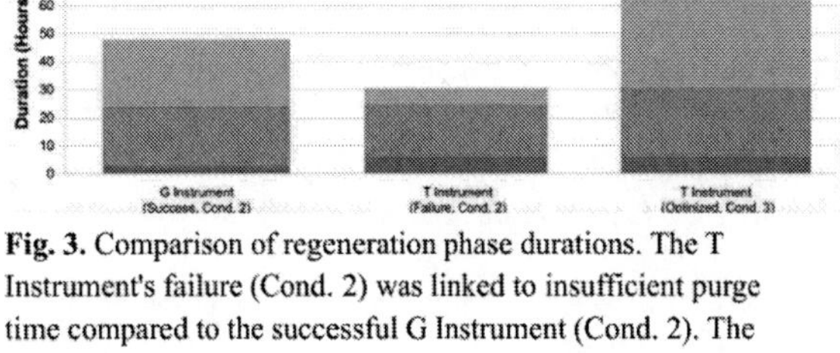

Fig. 3. Comparison of regeneration phase durations. The T Instrument's failure (Cond. 2) was linked to insufficient purge time compared to the successful G Instrument (Cond. 2). The

optimized protocol (Cond. 3) significantly extended the Heater and Standby purge times.

Table 2. Optimized O_2 Trap Regeneration Conditions for the T Instrument

Feature	**Inst. G**	**Inst. T**	**Inst. O (Ref)**
Status	OK (Stable since Feb 2025)	OK (Stable since Apr 2025)	Stable (Benchmark)
Regen. temp.	190°C	190°C	350°C
Regen. Gas (flow rate)	H2 + He (100/50 mL)	H2 + He (20/80 mL)	H2 + He (10/90 mL)
Regen. Time	3 hrs	6.5 hrs	6 hrs
Heater (Purge)	21 hrs (@ 190°C)	24 hrs (@ 190°C)	10 hrs (@ 350°C)
Standby	24 hrs (@ 150°C)	36–48 hrs (@ 150°C)	5 hrs
Total Cycle	~48 hrs	~3 to 4 Days	~24 hrs
Key Finding	Ghost peaks eliminated by extending purge from 12 hr to 21 hr.	Breakthrough managed by fixing injection limit to 55.	High Temp (350°C) allows fast cycles.

3.2 Quantitative Performance Improvement

The visual improvement was substantiated by quantitative data, as shown in Table 3. After implementing the optimized protocol, the peak area for argon recovered to its expected value, indicating a complete restoration of detector sensitivity. Furthermore, the baseline noise was reduced by over 95% to well below the acceptable limit of 20 μV, confirming the stability of the system.

Table 3. Instrument Performance Comparison

Parameter	**Before Regeneration**	**After Regeneration**
Ar peak area	38,000	39,000
Baseline noise (μV)	>500	<20

The peak areas for all target analytes increased by a factor of 40, demonstrating a complete recovery of detector sensitivity. Furthermore, the baseline noise was reduced by over 96%, and the resolution between the critical N_2/Ar pair was restored to a level well above the acceptable threshold.

3.3 Acceleration of the Troubleshooting Process

Perhaps the most significant outcome is the drastic reduction in the time required to resolve the issue. The traditional troubleshooting workflow for a similar problem has historically taken over a week in our lab. The AI-driven approach, from initial data input to the implementation of the final solution, was completed in just 16 hours. This represents a more than 85% reduction in instrument downtime, a critical advantage in a high-throughput analytical environment.

3.4 The Synergy of Human Expertise and AI

This case study powerfully illustrates the value of a human-AI collaborative approach. The AI's ability to rapidly process vast amounts of information and identify the underlying chemical principle across different proprietary systems was a significant advantage over manual research. However, the initial ambiguity of the AI's output underscores the continued importance of human expertise. The operator's ability to formulate a core hypothesis, critically evaluate the AI's suggestions, and provide targeted, refined inputs was the key to unlocking a fast and effective solution. This synergistic relationship transforms troubleshooting from a linear, time-consuming sequence into a dynamic and efficient problem-solving loop.

4 Conclusion

The successful resolution of O_2 trap breakthrough in a GC-HDID system using an AI-driven troubleshooting framework marks a significant advancement in analytical instrument maintenance. By integrating the "deep research" capabilities of an AI model with the critical domain knowledge of an experienced operator, we have demonstrated a methodology that is not only effective but also remarkably efficient.

This study has shown that:

1. An AI-powered approach can dramatically accelerate the root cause analysis of complex instrumentation failures.
2. The synergy between human expertise and AI is crucial for navigating the ambiguities and proprietary limitations often encountered in technical troubleshooting.
3. This methodology can reduce instrument downtime by over 85% compared to traditional methods, offering substantial benefits in terms of productivity and operational efficiency.

The framework presented here is not limited to O_2 trap issues or GC-HDID systems. It is a transferable and scalable model that can be adapted to a wide range of analytical instruments and failure modes. As AI models become more sophisticated, the adoption of such collaborative troubleshooting workflows will likely become a standard practice in scientific laboratories and industrial quality control environments, heralding a new era of intelligent and predictive instrument maintenance.

References

[1] D. R. Coulson, "The Helium Ionization Detector," *Journal of Gas Chromatography*, 3, 134-137 (1965).
[2] Restek Corporation, "A Technical Guide to Purifiers," *Restek ChromaBLOGraphy*, (2018).
[3] Oxygen Trap, PerkinElmer, n9301179, accessed on April 29, 2025.
[4] Agilent Technologies, *490 Micro GC Gas Analyzer Hardware User Manual*, (Agilent Technologies, Inc., 2010).
[5] US10428433B2 - Method to regenerate oxygen traps - Google Patents, accessed on April 29, 2025.
[6] EP3037151A1 - Method for regenerating an oxygen trap - Google Patents, accessed on April 29, 2025.
[7] Material Safety Data Sheet - PerkinElmer, n9301179, accessed on April 29, 2025.

 t pp s/di o.gpr5d(0g(0j (d: 3confd606) 1j 60100(

Indoor Air Quality monitoring using low-cost sensors: Experimental set-up and characterization procedure

Ramona Russo[1,2*], *Francesca* Rolle[1], *Francesca* Durbiano[1], *Francesca Romana* Pennecchi[1], *Stefano* Pavarelli [1], *Chiara* Festevole[3], and *Michela* Sega[1]

[1]INRiM, Istituto Nazionale di Ricerca Metrologica, 10135 Turin, Italy
[2]DENERG, Department of Energy "Galileo Ferraris", Politecnico di Torino, 10129 Turin, Italy
[3]Department of Chemistry, Università degli Studi di Torino, 10125 Turin, Italy

Abstract. Monitoring indoor environmental quality is a key requirement for addressing challenges related to smart cities and human health, including indoor air quality. Traditionally, indoor air monitoring has relied on high-end instrumentation, limiting large-scale and widespread deployment. The increasing availability of low-cost sensors offers new opportunities for scalable monitoring systems, provided their metrological performance is properly characterised. In this context, the Italian PRIN project MIRABLE (Measurement Infrastructure for Research on Healthy and Zero Energy Buildings in Novel Living Lab Ecosystems), involving the Italian National Metrology Institute (INRiM) and the Politecnico di Torino, aims to develop a multidomain measurement infrastructure for indoor environments using low-cost sensors in a full-scale living laboratory. At INRiM, a dedicated calibration system was developed to ensure the metrological traceability of CO_2 and CO measurements obtained from low-cost sensors. Reference gas mixtures were prepared gravimetrically according to the International Standard ISO 6142-1 and dynamically diluted to reach low concentration levels. Calibration was carried out using a primary non-dispersive infrared (NDIR) reference analyser and a specially designed calibration chamber. This study presents calibration results for selected CO_2 low-cost sensors and a preliminary evaluation of measurement uncertainty. The same methodology will be extended to low-cost CO and NO_x sensors in future work.

1 Introduction

Accurate assessment of indoor air quality (IAQ) has become a key public health priority, especially considering that individuals spend approximately 80 % of their time in confined environments such as homes, schools and offices [1]. Traditionally, IAQ monitoring and the assessment of ventilation rates have relied on high-end laboratory instruments, such as photoacoustic spectroscopy (PAS) systems; although extremely accurate, these instruments present significant limitations in terms of high cost, logistical complexity, and slow response times [2].

To overcome these barriers, there has been a rapid proliferation of monitoring systems based on low-cost sensor networks [3]. These systems offer a scalable alternative for achieving high spatial and temporal resolution, enabling the identification of local emission sources and the detection of transient events related to occupant activities that may not be captured by traditional instrumentation [2].

The integration of these sensors with Internet of Things (IoT) technologies enables the transition from sparse measurement points to truly distributed monitoring networks [3].

However, the use of low-cost sensors (LCSs) introduces significant metrological challenges. Low-cost sensors are often subject to drift, cross-sensitivity to gases other than those intended to be measured and dependence on thermo-hygrometric conditions (temperature and humidity) [4]. To ensure the robustness and reliability of the collected data, it is essential to implement rigorous procedures ranging from laboratory calibration to in-field validation.

In this context, metrological traceability represents the fundamental link between the deployment of low-cost sensors and the ability to support reliable decision-making for health protection and smart building management. It ensures that the measurements obtained are linked to international primary standards, such as reference gas mixtures prepared in accordance with the International Standard ISO 6145-1 [5], thereby guaranteeing that the reported concentrations are accurate and not affected by systematic errors. Only through traceable measurement results is it possible to correctly quantify personal exposure to pollutants, enabling both authorities and citizens to take informed actions to mitigate health risks [3]. In modern buildings, defined as "cognitive", the metrological characterisation of sensor performance is not merely an observational tool but a critical prerequisite for automation, control strategies, and the large-scale deployment of measurement infrastructures in complex urban environments [3].

* Corresponding author: r.russo@inrim.it

2 MIRABLE Project

The "Measurement Infrastructure for Research on heAlthy and zero energy Buildings in novel Living lab Ecosystems" (MIRABLE) project, already presented in [6], aims to develop an integrated system for measuring environmental parameters within a living lab. The system development is divided into two phases. The first phase, currently at a more advanced stage, involves the creation of a desktop multisensor for measuring parameters such as temperature, relative humidity and other environmental indicators. The system also includes a component integrated into the chair, designed to measure temperature and relative humidity at two points: on the backrest, at neck height, and on the lower support of the chair, near the feet. The second phase involves the development of an instrumented living lab, in which the entire structure integrates fixed sensors distributed in different positions to monitor the main environmental parameters. In order to obtain more accurate control of the factors that influence the occupants' comfort, the functioning of some wearable sensors already available on the market, capable of measuring parameters such as temperature, physiological variables and lighting, was also analysed. Finally, this article presents the carbon dioxide (CO_2) concentration measurements acquired by using low-cost sensors, which are used to define the generic calibration procedure for low-cost sensors to be applied throughout the project.

3 Analysis of CO_2 amount fraction with low-cost sensors

The analyses conducted at INRiM involved three low-cost sensors and two high-quality sensors for CO_2 measurement.

3.1 Materials and Method

To ensure metrological traceability, in accordance with ISO 17025 and ISO 17043 International Standards [7], [8], the tests conducted at INRiM were performed using a high-quality CO_2 analyser as a reference instrument, previously calibrated according to the Institute's procedures using certified gas mixtures.

Specifically, the instrument used is a LI-cor 850, manufactured by LI-cor, based on NDIR (Non-Dispersive Infrared Technology), with a measurement range between 0 ppm and 20000 ppm and a declared accuracy of ± 1,5% of the reading [9].

In this article, this instrument is referred to as the primary instrument and identified as LI-cor 1.

A second high-quality instrument was also used, also a LI-cor 850 with identical nominal characteristics to the first, referred to as the secondary instrument (LI-cor 2). This instrument was not previously calibrated, but it was used to evaluate the typical behaviour of a high-quality uncalibrated sensor in comparison with the low-cost ones.

Three low-cost sensors were used, identified as LCS 1, LCS 2 and LCS 3. These are based on NDIR technology for CO_2 detection.
According to the manufacturer's datasheet, the declared accuracy varies depending on the measurement range:

- (0-1250) ppm: ± 1 % FS (± 50 ppm)
- (1250-2500) ppm: ± 2 % FS (±1 00 ppm)
- (2500-5000) ppm: ± 5 % FS (± 250 ppm)

The data was acquired at a sampling rate of one data point every 5 s.

To test the low-cost CO_2 sensors, which were not designed for use with a dedicated gas mixture inlet, an insulator measuring 80x80x80 cm, previously characterised, was designed at INRiM (Fig.1). The isolator was used for the simultaneous positioning of all the sensors used in the study, namely the two LI-cor 850 and the three low-cost sensors and was used as a mixing chamber. To ensure the homogeneous distribution of gases inside the isolator, a 15 V fan was used.

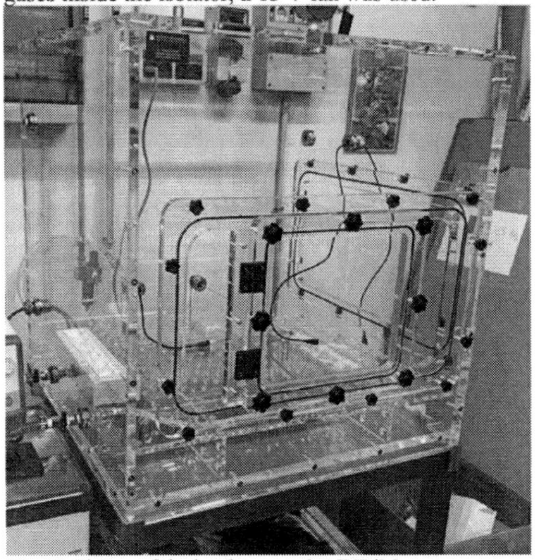

Fig. 1. Picture of the insulator used at INRiM

Two gas mixtures of CO_2 were employed: the first is CO_2 in synthetic air (SA) with a nominal concentration of 500.3 ppm (SAPIO s/n 5113566) and the second CO_2 in nitrogen (N_2) with a concentration of 80325 ppm (INRiM s/n D580464). A cylinder of pure N_2, (grade 6.0) supplied by Air Liquide (Italy) was also used as zero gas. The 500.3 ppm CO_2 mixture was used to define the initial and final span of the two LI-cor 850 instruments, while the 80325 ppm mixture was used for dynamic mixing inside the isolator.

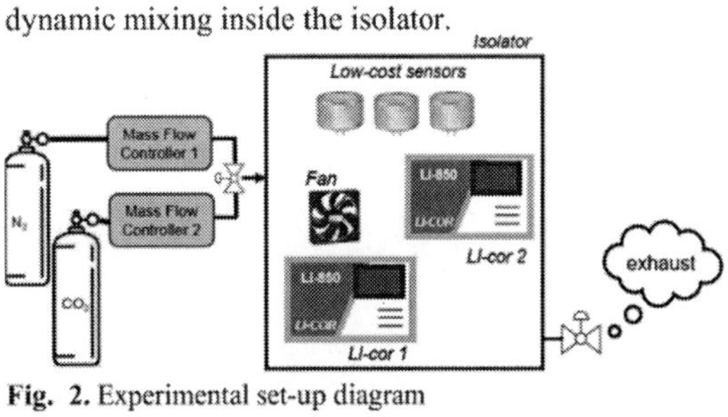

Fig. 2. Experimental set-up diagram

Knowing the volume of the isolator, the opening time of the CO_2 cylinder at 80325 ppm required to reach a target concentration inside the chamber was estimated. The gas flow was controlled by two calibrated Mass Flow Controllers (MFC) (MKS Instruments, USA), with full scales of 1000 standard cubic centimetres per minute (SCCM) and 2000 SCCM, respectively.

The experimental approach involved a series of measurements at different CO_2 concentrations, aimed at evaluating the response time of the sensors, their short-term stability and evaluating the uncertainty associated with the readings of the low-cost sensors. These analyses were carried out by normalising the measurements, using the measurements provided by the high-quality calibrated LI-cor 1 instrument as a reference.

For each test, the average measured value (avg. v.) was considered, calculated by excluding the extreme values of the measurement, which could potentially be affected by temporary instability. The average difference avg(Δ) value was also calculated, defined as the difference between the average value measured by the sensor in question and the average reference value (avg. ref. v.).

$$\text{avg}(\Delta) = \text{avg. v.} - \text{avg. ref. v.} \tag{1}$$

The standard deviation of the measurement differences $s(\Delta)$ was analysed, and the expanded repeatability uncertainty $U(\Delta)$ and the relative expanded uncertainty $U_r(\Delta)$, evaluated as explained below in equations (2) and (3), were considered relevant. A coverage factor k of 2 was adopted.

$$U(\Delta) = k \cdot s(\Delta) \tag{2}$$
$$U_r(\Delta) = U(\Delta)/\text{avg}(\Delta) \tag{3}$$

3.2 Experiment and results

The experiment began once the isolator had been closed. The initial phase involved continuously flushing N_2 into the chamber for 4 h. During this process, the exit valve was used to manage gas outflow and prevent overpressure conditions that could have compromised the sensors or the chamber's structural integrity. Internal overpressure was monitored throughout this phase to ensure it never exceeded 40 mbar, enabling the system to reach the minimum CO_2 concentration level of 5 ppm, detected by the calibrated reference instrument. At this concentration, the LI-cor 2 provided measurements that were closely comparable to the reference instrument, recording an average value of 2 ppm; in contrast, the low-cost CO_2 sensors exhibited significant deviations, with respective mean biases of 250 ppm, 188 ppm and 210 ppm (Fig. 3), respectively. These results highlight a pronounced systematic offset even at the lower detection limits.

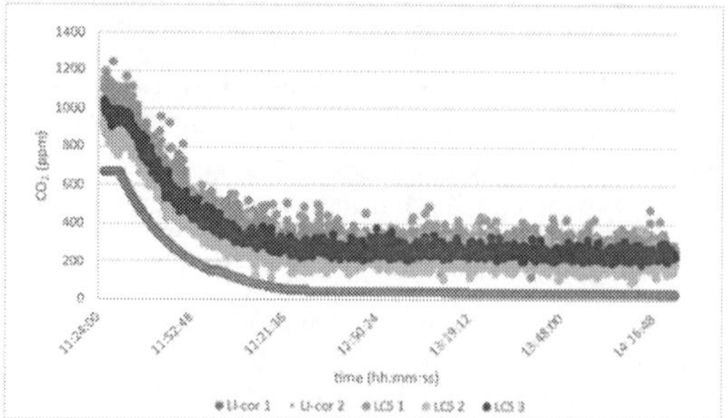

Fig. 3. Measurement of CO_2 from ambient concentration to nearly zero concentration.

Various concentrations of CO_2 were analysed, namely: 5, 400, 840, 1360, 3000, 4000 and 5000 ppm. Not all levels were analysed in detail, as some fell outside the calibration range of the reference instrument. However, the overall trend of the measurements taken by the low-cost sensors is consistent with that of the reference instruments (Fig. 4).

The response time of the low-cost sensors was approximately 2 min, which is comparable to that of the high-quality instruments used in the test (Fig. 4).

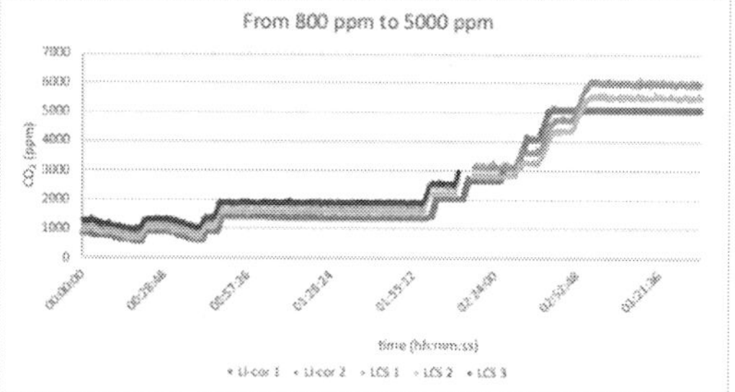

Fig. 4. Measurements taken from concentrations of 800 ppm to 5000 ppm of CO_2.

Figure 4 shows that, at a concentration of 3000 ppm, the LCS 3 low-cost sensor stopped collecting data. The analyses had been ongoing for three consecutive days after the LCS 3 sensor failed; its data collection was suspended to prevent further data loss. In the range between 3000 ppm and 4000 ppm, the low-cost sensors consistently underestimated CO_2 levels, probably due to their recent reactivation.

Table 1 shows the analysed data relating to concentrations of 5, 400, 840 and 1360 ppm. The high-quality sensors (LI-cor 2) show low extended repeatability uncertainties $U(\Delta)$ (< 4 ppm), while the low-cost sensors show much higher values (up to ~ 80–90 ppm), exceeding the accuracy declared by the manufacturer (± 50 ppm, 1 % FS). The relative expanded uncertainty, which for low-cost sensors reaches values of around 20–60 %, makes measurements unreliable without an adequate metrological procedure, especially at low concentrations.

Table 1. Comparative analysis of data collected by sensors for different levels of CO_2 mole fraction (5, 400, 840, 1360 ppm).

LI-cor 1 measurement: 5 ppm				
	LI-cor 2	LCS 1	LCS 2	LCS 3
avg. v.	2 ppm	250 ppm	188 ppm	210 ppm
avg(Δ)	-3	245	183	205
s(Δ)	0.243	44.367	40.046	21.847
U(Δ)	0.486	88.734	80.092	43.695
*U*r(Δ)	16 %	36 %	44 %	21 %
LI-cor 1 measurement: 400 ppm				
avg. v.	398 ppm	695 ppm	537 ppm	635 ppm
avg(Δ)	-2	295	137	235
s(Δ)	0.417	43.461	40.752	24.604
U(Δ)	0.835	86.923	81.504	49.207
*U*r(Δ)	35 %	29 %	60 %	21 %
LI-cor 1 measurement: 840 ppm				
avg. v.	840 ppm	1219 ppm	1008 ppm	120 ppm
avg(Δ)	0	379	169	280
s(Δ)	0.673	36.876	32.245	26.396
U(Δ)	1.345	73.751	64.489	52.792
*U*r(Δ)	*	19 %	38 %	19 %
LI-cor 1 measurement: 1360 ppm				
avg. v.	1364 ppm	1816 ppm	1579 ppm	1888 ppm
avg(Δ)	4	456	220	529
s(Δ)	0.758	38.036	32.364	24.298
U(Δ)	1.515	76.072	64.728	48.596
*U*r(Δ)	37 %	17 %	29 %	9 %

*Undetermined value because the average difference is zero

Figures 5 and 6 below shows, as an example, the analysis of the mole fraction of CO_2 measured with low-cost sensors, normalised to a reference value of 400 ppm. Figure 5 shows the absolute values measured over time. Even under steady-state conditions, low-cost sensors show significant systematic offsets from the reference, with average values between + 137 ppm and + 295 ppm, indicating poor absolute accuracy. Figure 6 shows the effect of correction to the reference value (400 ppm). Normalization reduces the average offset, but does not eliminate data dispersion, which remains high for low-cost sensors, indicating that the error is not only systematic but also related to repeatability.

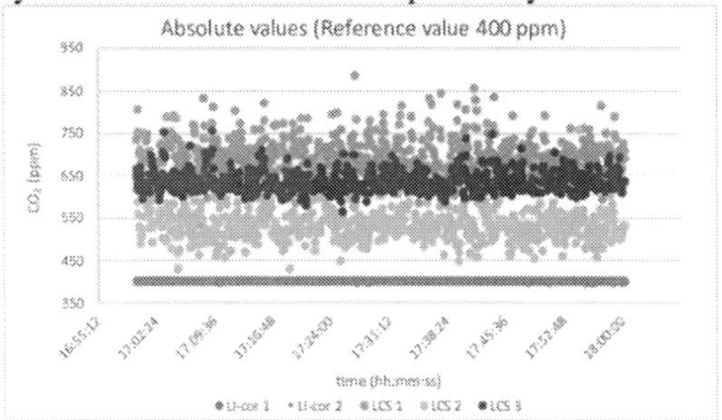

Fig. 5. Absolute CO_2 values measured by all sensors

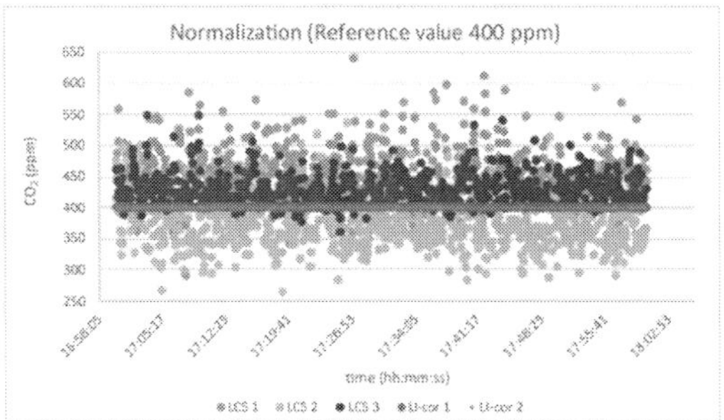

Fig. 6. Data normalised to the reference value of 400 ppm of CO_2

This demonstrates that normalisation to a single reference point is not sufficient to guarantee reliable measurements with low-cost sensors, and that multi-point calibration procedures and structured metrological analyses are necessary.

It is interesting to note that the deviation between the measured value and the reference value (avg(Δ)) increases as the concentration increases (Fig. 7); on the contrary, the standard deviation of the difference, and consequently the standard uncertainty of the correction, tends to decrease as the concentration increases (Fig. 8).

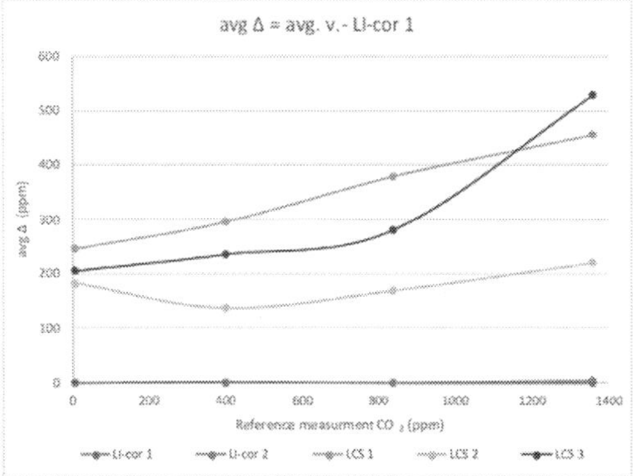

Fig. 7. Difference between measured value and reference value

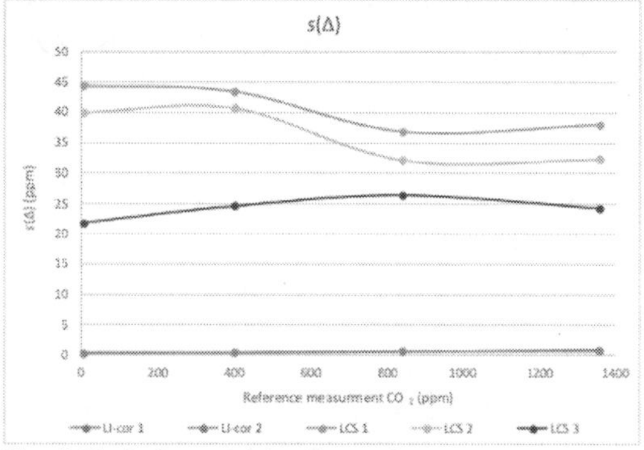

Fig. 8. Typical uncertainty of correction

Fig. 9 shows the absolute values measured by the different sensors as a function of the reference CO_2 concentration, highlighting increasing systematic deviations and dispersion that increases with concentration, particularly for low-cost sensors (LCS 1–3).

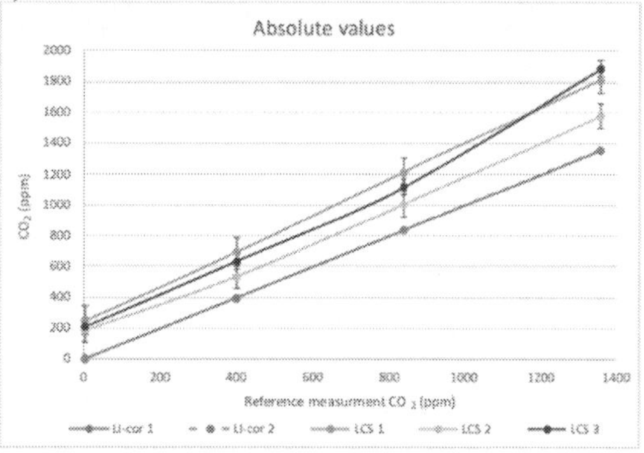

Fig. 9. Absolute CO_2 values measured in the range 5-1360 ppm, reported with repeatability expanded

4 Conclusion

This paper presents the experimental set-up and characterization procedure for a preliminary analysis of the measurement uncertainty of the mole fraction of CO_2 obtained with low-cost sensors, compared with reference instruments (LI-cor) in a range between 5 ppm and 1360 ppm. While the reference instrumentation guarantees low uncertainties (< 4 ppm), the low-cost sensors show dispersions of up to 80–90 ppm, exceeding the specifications declared by the manufacturer. The analysis highlights the need for calibration, as the direct use of these sensors, without corrections, leads to high and uncontrolled errors. Metrological traceability is therefore essential to ensure data reliability. Although low-cost sensors are an affordable solution for indoor air quality monitoring, the validity of their measurements depends heavily on the adoption of standardised calibration protocols. This study mainly analysed repeatability uncertainty; however, in view of future developments, other components of the uncertainty budget will also need to be considered. The obtained results provide the basis for defining a rigorous calibration procedure specific to low-cost CO_2 sensors.

Acknowledgment

The MIRABLE Project was funded by the Italian PIANO NAZIONALE DI RIPRESA E RESILIENZA (PNRR) - Missione 4 "Istruzione e Ricerca" - Componente C2 Investimento 1.1, "Fondo per il Programma Nazionale di Ricerca e Progetti di Rilevante Interesse Nazionale (PRIN)" Decreto Direttoriale n. 104 del 2 febbraio 2022. Progetto PRIN 2022 «MIRABLE – Measurement Infrastructure for Research on heAlthy and zero energy Buildings in novel Living lab Ecosystems» Codice di progetto: 2022ZNJHHY.

References

[1] J. Wu, J. Weng, B. Xia, Y. Zhao, and Q. Song, 'The Synergistic Effect of PM2.5 and CO2 Concentrations on Occupant Satisfaction and Work Productivity in a Meeting Room', *IJERPH*, vol. 18, no. 8, Art. no. 8 (2021). doi: 10.3390/ijerph18084109.

[2] V. Gentile and M. Perino, 'Low-Cost Tracer Gas Sensor for Building Ventilation', *Journal of Physics: Conference Series*, vol. 3140, p. 092001 (2025). doi: 10.1088/1742-6596/3140/9/092001.

[3] S. Lopes *et al.*, 'Low-Cost Sensor Systems and IoT Technologies for Indoor Air Quality Monitoring: Instrumentation, Models, Implementation, and Perspectives for Validation', *Sensors*, vol. 25, p. 7567 (2025). doi: 10.3390/s25247567.

[4] E. Bagkis, T. Kassandros, and K. Karatzas, 'Learning Calibration Functions on the Fly: Hybrid Batch Online Stacking Ensembles for the Calibration of Low-Cost Air Quality Sensor Networks in the Presence of Concept Drift', *Atmosphere*, vol. 13, no. 3, p. 416 (2022). doi: 10.3390/atmos13030416.

[5] 'ISO 6142-1:2015', ISO. Accessed: Jan. 12, 2026. [Online]. Available: https://www.iso.org/standard/59631.html

[6] F. Rolle *et al.*, 'Metrological characterisation of low-cost and wearable sensors for research on healthy buildings in a novel Living Laboratory', *EPJ Web of Conferences*, vol. 323 (2025). doi: 10.1051/epjconf/202532304001.

[7] 'ISO - ISO/IEC 17025 — Testing and calibration laboratories', ISO. Accessed: Apr. 17, 2025. [Online]. Available: https://www.iso.org/ISO-IEC-17025-testing-and-calibration-laboratories.html

[8] 'ISO/IEC 17043:2023', ISO. Accessed: Jan. 08, 2026. [Online]. Available: https://www.iso.org/standard/80864.html

[9] 'LI-COR Environmental', LI-COR Environmental. Accessed: Jan. 08, 2026. [Online]. Available: https://www.licor.com/products/gas-analysis/LI-830-LI-850

Advances in the production of traceable amount fraction and isotope ratio gas reference materials to underpin climate monitoring

Ruth Hill-Pearce [1*], *Emily* Hopkinson[1], *Aimee* Hillier[1], *Sarah*, Channell[1], *Freya* Wilson[1], *Isabell* Von Rein[1], *Christoph* Nehrbass-Ahles[1], *Tom* Gardiner[1] and *Paul* Brewer[1].

[1]National Physical Laboratory, Hampton Road, Teddington, TW11 0LW, U.K.

Abstract. The National Physical Laboratory (NPL) in the United Kingdom provides traceable gas reference materials for calibration of atmospheric gas analysers where low uncertainty measurements of the amount fraction and/or stable isotope ratio of atmospheric components are required. Primary gas Reference Materials (PRMs) are available for amount fraction traceable to the SI and, for CO_2 amount fractions between 380 and 800 µmol mol^{-1}, can be linked to the WMO-CO_2-X2019 scale via the BIPM.QM-P5 comparison. Traceable stable isotope ratio PRMs for the most potent anthropogenic greenhouse gases and climate relevant species including carbon dioxide (CO_2), methane (CH_4), nitrous oxide (N_2O) and carbon monoxide (CO) are available and are traceable to current globally recognised isotope reference scales including VPDB for $\delta^{13}C$-CO_2, $\delta^{18}O$-CO_2 and $\delta^{13}C$-CH_4, VSMOW/SLAP for $\delta^{2}H$-CH_4 and $\delta^{18}O$-N_2O and Air N_2 for $\delta^{15}N$-N_2O. This paper details recent advances in the production of PRMs at NPL and outlines the PRMs and services currently being developed.

1 Traceability for primary gas reference materials (PRMs) for greenhouse gas amount fraction.

National Metrology Institutes (NMIs) as well as Designated Institutes (DIs) maintain and develop the highest-level primary measurement standards. For amount fraction of gases, these are typically primary reference gas materials (PRMs), prepared by precise gravimetry, providing traceability to the International System of Units (SI) [1]. The preparation and certification of PRMs establishes an unbroken, documented chain of comparisons, ensuring measurement results are accurate, comparable, and internationally recognised [2]. The NMIs/DIs then disseminate this traceability to end-users and secondary calibration laboratories.

The Bureau International des Poids et Mesures (BIPM) and NMIs/DIs organise intercomparions. These key comparisons are crucial to ensure global consistency and traceability to the SI providing a method for NMIs to benchmark their capabilities in preparing and analysing gas reference materials [3]. Through comparing results across multiple laboratories systematic biases can be identified and corrected for, reported uncertainties can be verified and consistency in value assignment is improved [3].

International key comparisons including CCQM-K120 for CO_2 in air [4], CCQM-K82 for methane in air [5] and CCQM-K68 for N_2O in air [6], have played a critical role in enabling NMIs/DIs to demonstrate a reduction in uncertainties in the amount fraction of gas mixtures towards those required by the user community. Achieved uncertainties are documented in the BIPM Key Comparison Database (KCDB) [7].

Due to the need to measure small changes in the amount fraction of greenhouse gases in the atmosphere, global greenhouse gas monitoring requires precise and comparable data between monitoring stations if the data is to be useful for modelling and flux studies [8]. The World Metrological Organisation Global Atmospheric Watch (WMO-GAW) has published inter-laboratory compatibility goals which define the maximum bias for these measurements see Table 1 [8]. The requirements for the uncertainty of gas reference materials for global greenhouse gas monitoring are ideally an order of magnitude lower than the compatibility goals to enable monitoring stations to achieve the compatibility goals.

Despite the consistent reductions in uncertainty achieved [9,10] and documented in key comparisons [4-6], the extremely low uncertainties required continue to be challenging to achieve using the SI approach which evaluates the total uncertainty in amount fraction [13]. For gases such as CO_2, where reference materials for instrument calibration which are SI traceable currently does not provide sufficiently low uncertainty to meet the goals, the CO_2 reference gas materials scale system enables the production and value assignment of gas reference materials which have very small uncertainties in relationship to each other [13]. This low relative uncertainty leads to a high degree of internal consistency which is crucial for those atmospheric monitoring

* Corresponding author: ruth.pearce@npl.co.uk

applications where small temporal and spatial changes are important. [8]

Table 1. Inter-laboratory compatibility goals published by the WMO-GAW [8].

Component	Compatibility goal	Extended compatibility goal	Range in unpolluted troposphere	Range covered by the WMO scale
CO_2	± 0.1 ppm (Northern hemisphere) ± 0.05 ppm (South. hemisphere)	± 0.2 ppm	360 - 450 ppm	250 – 520 ppm
CH_4	± 2 ppb	± 5 ppb	1700 – 2100 ppb	300 – 2600 ppb
CO	± 2 ppb	± 5 ppb	30 – 300 ppb	20 -500 ppb
N_2O	± 0.1 ppb	± 0.3 ppb	320 – 335 ppb	260 – 370 ppb
SF_6	± 0.02 ppt	± 0.05 ppt	6 – 10 ppt	1.1 – 9.8 ppt
H_2	± 2 ppb	± 5 ppb	450 – 600 ppb	140 –1200 ppb
$\delta^{13}C\text{-}CO_2$	± 0.01‰	± 0.1‰	-7.5 to -9‰ vs. VPDB	
$\delta^{18}O\text{-}CO_2$	± 0.05‰	± 0.1‰	-2 to +2‰ vs. VPDB	
$\Delta^{14}C\text{-}CO_2$	± 0.5‰	± 3‰	0-70‰	
$\Delta^{14}C\text{-}CH_4$	± 0.5‰		50-350‰	
$\Delta^{14}C\text{-}CO$	± 2 molecules cm^{-3}		0-25 molecules cm^{-3}	
$\delta^{13}C\text{-}CH_4$	± 0.02‰	± 0.2‰		
$\delta D\text{-}CH_4$	± 1‰	± 5‰		
O_2/N_2	± 2 per meg	± 10 per meg	-250 to -800 per meg (vs. SIO scale)	

The amount fraction of CO_2 in the atmosphere is traceable to the WMO-CO_2-X2019 CO_2 in air scale provided by the WMO [11]. The NOAA Global Monitoring Laboratory currently serves as the WMO-GAW Central Calibration Laboratory (CCL) for CO_2 and is responsible for maintaining the scale [11]. This scale has a reported consistency between tertiary standards of the order of 0.01 µmol mol^{-1}, with the SI traceability of their primary standards demonstrated in the CCQM-K120 comparison [11]. This relationship between reference materials in the scale approach is maintained in a database, and the value of any reference material can be traced back to the primary standard set defining the scale [11].

To reduce the burden on one laboratory, several NMIs/DIs are working toward providing scale traceable reference materials. The requirement for inter-compatible scales is that differences in scales must be close to the internal consistencies which are currently achieved by the WMO scale approach [12]. To minimise any potential scale bias, it is planned to maintain scale equivalence equations by using CIPM MRA on-going key comparisons [12, 8]. The comparisons of CO_2 in air scales (BIPM.QM-P5) will ensure consistency between the measurements traceable to the scales at the 0.01 µmol mol^{-1} level [12].

The UK's National Physical Laboratory (NPL) has prepared the NPL 2023_CO_2 scale and tertiary reference materials traceable to this scale are currently being measured as part of BIPM.QM-P5 pilot comparison [12] which will provide algorithms to link amount fractions on each participating scale to the NOAA X2019 CO_2 scale. Thus, allowing participating NMIs, such as NPL, to provide PRMs and certified reference materials (CRMs) to the WMO monitoring community [13].

2 Traceability of PRMs for stable isotope ratio in greenhouse gases.

The measurement of the ratio of isotopes in atmospheric gases can be a powerful tool for investigating environmental processes including sources and sinks of greenhouse gases [14]. The large-scale adoption of optical laser-based isotope ratio measurement techniques at greenhouse gas monitoring stations allow the apportionment of greenhouse gas fluxes to emission sectors [14]. However, traceable PRMs or CRMs for calibration of optical isotope ratio analysers are limited.in availability.

Under the CIPM Mutual Recognition Arrangement (CIPM MRA), a specific traceability exception exists for delta value isotope ratio measurements [15]. This exception allows measurements of isotope ratio to be made traceable to artefacts recognised as International Standards rather than directly to the SI. This is because, in the case of stable isotope ratio in gases, artefact-based scales currently provide lower measurement precision than existing SI-traceable methods [13]. Absolute isotope ratio measurements traceable to the SI [16, 17] have not yet achieved the desired low uncertainty required for greenhouse gas monitoring.

The internationally recognised consensus materials for stable isotope ratio relevant to greenhouse gases [18] are the Vienna Peedee Belemnite (VPDB) scale, typically realised through carbonate materials such as the now depleted NBS19 which is used for measurements of $\delta^{13}C\text{-}CO_2$ and $\delta^{13}C\text{-}CH_4$ and $\delta^{18}O\text{-}CO_2$; the Vienna Standard Mean Ocean Water and Standard Light Antarctic Precipitation scales (VSMOW/SLAP) for $\delta^{2}H\text{-}CH_4$ and $\delta^{18}O\text{-}N_2O$; and $AirN_2$ for $\delta^{15}N\text{-}N_2O$. The CCQM Isotope Ratio Working Group (IRWG) [19], established in 2018, is the primary body responsible for advancing traceability capabilities for stable isotope measurement and moving toward absolute SI-traceable ratios. They work in close collaboration with the Gas Analyses Working Group (GAWG) and have co-founded a new task group on Carbon Dioxide and Methane Stable Isotope Ratio Measurements (CCQM-GAWG-IRWG-TG-ISOTOP) established in 2023, to facilitate building a robust metrological support infrastructure for the measurement of stable isotope ratios in atmospheric greenhouse gases [20]. Worldwide consistency and traceability in measurements of isotope ratio in gases is challenging which can lead to differences in reported values of PRMs/CRMs used for calibration depending on which scale and institute the PRM/CRM producer is traceable to [21]. The BIPM stable isotopes of CO_2 pilot study (CCQM-P204) [22] evaluated the ability of different laboratories to accurately measure the isotope ratios in pure CO_2 gas

https://doi.org/50.50j 5/ep3conf/(0(61j (0100(

samples. This study highlighted the difficulties with primary realisation methods of the scale and the need for improved reference materials and consistency in methods and harmonised measurement practices. While no single reference value was calculated due to differences in the scales used by the participants, the study revealed offsets between participants which significantly exceed the WMO compatibility targets of 0.01‰ and 0.05‰ for $\delta^{13}C$-CO_2 and $\delta^{18}O$-CO_2 respectively. The challenges in achieving comparable measurements of the isotopic ratio of pure CO_2 is a fundamental challenge in the production of compatible diluted PRMs/CRMs.

Following on from CCQM P204, the ongoing pilot study CCQM-P239 [23] for CO_2 in air or N_2, provided samples containing CO_2 in the range of 380 µmol mol^{-1} to 800 µmol mol^{-1} and $\delta^{13}C$ and $\delta^{18}O$-CO_2 values from 1‰ to -43‰ and -7‰ to -35‰, respectively to participants for measurement. Precise and comparable measurements of the isotope ratio of greenhouse gases in air are vital to studies of fractionation on dilution during the production of ambient amount fraction isotope ratio PRMs/CRMs

3 Recent developments in PRMs for amount fraction.

3.1 The importance of matrix gases

Reductions in the SI traceable uncertainty offered by NMIs for PRMs produced at the low amount fractions at which greenhouse gases exist in the atmosphere have primarily been driven by improvements to gravimetric preparation techniques and impurity analysis of matrix gases. Analysis of the trace amount fractions of greenhouse gases present in the Ar, O_2 and N_2 which comprise the matrix is challenging due to the very low amount fractions but is critical for an accurate analysis of the final amount fraction in the PRM as the amounts can be significant due to the low amount fractions at which these PRMs are required.

Measurements of impurities in matrix gases are not only hampered by the limit of detection of analysers but also by the lack of traceable reference materials at trace level. Dynamic dilution [24] offers a practical methodology for producing gas reference materials at very low levels where adsorption effects present a significant issue. Preconcentration has also been developed at NPL [25] to enable lower limits of detection for purity analysis of matrix gases [10, 26] where the amount fraction of one or more isotopocules is often very low.

Collisions of the measured gas with matrix gases leads to changes in the adsorption profile due to pressure broadening [27]. Changing the matrix composition away from that of whole air can lead to biases in the reported amount fraction due to changes in the amount of pressure broadening which can significantly affect the reported amount fractions. The BIPM has published tolerances for matrix amount fraction [28] and also within [4-6], within which changes to pressure broadening are calculated to be negligible. NPL produces PRMs in air with amount fractions of Ar, O_2 and N_2 within the stated tolerances. The uncertainties offered by NPL for PRMs of CO_2, CH_4 and N_2O in air are stated in table 2 and discussed in section 3.3

3.2 The NPL-2023-CO_2 scale

The NPL CO_2 scale comprises a suite of nine 'super primaries' with traceability to the SI, the NPL primary scale and the NPL secondary scale each with nine reference materials with amount fractions spanning 380 µmol mol^{-1} to 800 µmol mol^{-1}. Tertiary reference materials are compared to the secondary scale and

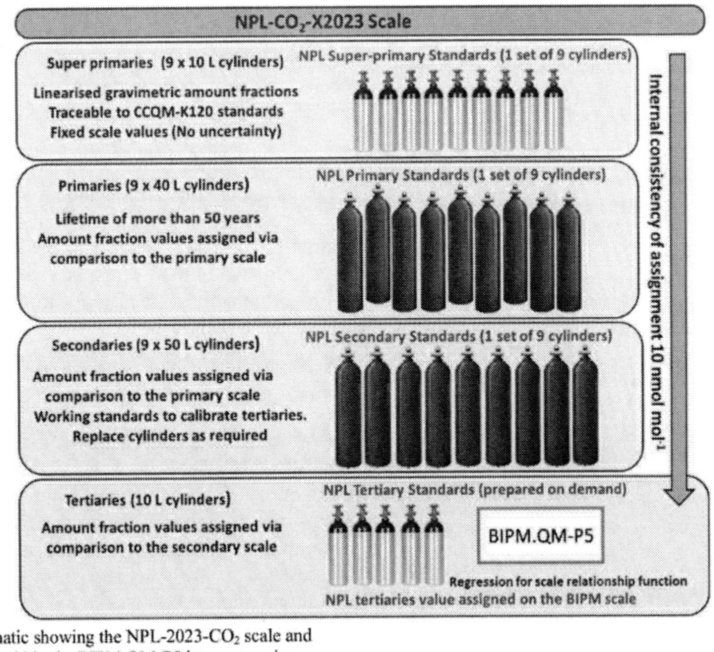

Fig. 1. Schematic showing the NPL-2023-CO_2 scale and comparisons within the BIPM.QM-P5 intercomparison.

 https://doi.org/50.50j 5/ep3conf/(0(61j (0100(

distributed to monitoring stations for instrument calibration. Fig. 1. outlines the NPL CO_2 scale and the scale comparison BIPM.QM-P5 [29].

The super primary scale is traceable to the SI via comparison with reference materials evaluated in the CCQM K120 key comparison which span the amount fraction range of the CO_2 scale. The super primaries are produced by exchange dilution [30] to mitigate adsorption effects on the cylinder walls. The gravimetric values of the super primaries provide traceability for the NPL-2023-CO_2 scale via sequential certification of values for the primary and secondary scales.

The reproducibility of assignment of values to a suite of tertiary reference materials will be monitored through time to ascertain final values for scale transfer uncertainties. Stability of the suites of gas reference materials and the individual reference materials comprising each suite will be measured via the annual comparisons. Changes to the calibration function of each suite of reference materials may remain comparatively consistent even if drift in a single reference material occurs. The stability of the individual reference materials will be monitored through time by comparison of the residuals from the fitted function of the suite of reference materials.

3.3 Amount fraction greenhouse gas PRMs and CGMs currently available from NPL

NPL provides bespoke PRMs and certified gas reference materials (CGMs). Gas reference materials produced at the NPL are supplied in a synthetic air matrix comprising argon, oxygen and nitrogen at ambient amount fractions. These reference materials are suitable for calibrating a range of atmospheric gas analysers. Table 2. shows the PRMs and uncertainties currently available from NPL for ambient amount fraction greenhouse gases.

Component	Amount fraction range [µmol mol^{-1}]	Typical amount fraction uncertainty (k=2) [%]	Matrix
CO_2*	380 - 800 **	0.2	Synthetic air. N_2 available on request
CH_4	1.8 - 2.5	0.2	Synthetic air. N_2 available on request
N_2O	0.32 – 0.35	0.5	Synthetic air. N_2 available on request

Table 2. SI traceable amount fraction greenhouse gas PRMs currently available from the NPL. * CO_2 gas reference materials traceable to the NPL 2023 CO_2 scale are expected to become available during 2026. Gas reference materials can be ordered and enquiries made by e-mailing gas_enquiries@npl.co.uk** Other amount fractions are available on request.

4 Recent development of PRMs/CRMs for isotope ratio.

Isotope ratio reference materials have been developed for calibration of optical isotope ratio analysers for the three most potent anthropogenic greenhouses: CO_2, CH_4 and N_2O. These isotopic PRMs are prepared from large volumes of natural pure source gases and have isotope ratios characterised by IRMS with traceability to VPDB for $\delta^{13}C$-CO_2, $\delta^{13}C$-CH_4 and $\delta^{18}O$-CO_2, VSMOW/SLAP for $\delta^{2}H$ in CH_4 and $\delta^{18}O$ in N_2O, and AirN_2 for $\delta^{15}N$-N_2O. The source gases were chosen to span a range of isotope ratios relevant for source attribution.

CO_2 sources cover the $\delta^{13}C$-CO_2 range of nominally + 2‰ to -42 ‰ with corresponding $\delta^{18}O$-CO_2 values of -8 ‰ to -36 ‰ [29]. Blends of CO_2 were produced to target specific $\delta^{13}C$ values. The tuning of the $\delta^{18}O$ CO_2 using enriched water has been demonstrated [31]. Further sources of CO_2 and CH_4 including biogenic sources are currently being evaluated for the purity and volume requirements required for long term supply of PRMs. Biogenic sources will enable the production of PRMs with certified $\delta^{14}C$ ratios. Pure N_2O reference materials were obtained from Empa and match the isotope ratios of the reference materials marked RM1b, RM3a, RM3b and RM5 in J. Mohn *et al.* [32]. Further N_2O isotope ratios and larger volumes of reference materials are currently being developed in-house.

The pure source gases are diluted to a range of near ambient amount fractions in an air matrix comprising Ar, O_2 and N_2 at ambient amount fractions to avoid biases when measuring with whole or synthetic air on optical isotope ratio analysers due to pressure broadening. The isotope ratios of the pure and ambient reference materials are measured by Isotope Ratio Mass Spectrometry (IRMS) to investigate potential fractionation on dilution and to provide traceability to existing isotope ratio scales [31].

Further PRMs are measured in house using Optical Isotope Ratio Spectroscopy (OIRS) to assign isotope ratios and ensure consistency between newly produced diluted PRMs and those measured by IRMS. Calibration protocols have been developed for OIRS to enable traceable calibration using PRMs with certified isotope ratios alongside uncertainty evaluations for the calibration procedures.

Multi component PRMs can be produced with certified isotope ratios for CO_2, CH_4 and N_2O. Additional components are added as required to address the sensitivities of the measurement techniques, for example, Kr and CO have been added to PRMs for measurements of the isotope ratio of CO_2 by IRMS and for measurements of the isotope ratio of N_2O by OIRS.

4.1 Large scale production of CH_4 isotope ratio PRMs.

A recent comparison study which formed part of the EMPIR 21GRD04 isoMET project required the production of over 30 calibration reference materials with a range of near ambient $\delta^{13}C$-CH_4 isotope ratios for a comparison of measurement compatibility at seven institutes. The institutes used a range of isotope ratio measurement instruments including IRMS and OIRS with and without preconcentration. Harmonised measurement protocols were developed [33]. Each partner received a suite of calibration PRMs to ensure traceability to VPDB and VSMOW/SLAP. The calibration PRMs were gravimetrically produced from four isotopically characterised fossil CH_4 sources with near ambient $\delta^{13}C$. Where requested, ambient amount fraction CO_2 and N_2O were added as the presence of these gases may cause changes to the reported isotope ratio measurements for OIRS techniques. Higher amount fraction CH_4 in N_2 reference materials (500 - 600 μmol mol^{-1}) were produced for calibration of OIRS instruments which utilised preconcentration of ambient amount fraction samples. Ambient amount fraction PRMs were produced as travelling standards which were then measured at each institute. Results of the comparison are being prepared for publication [34]

4.2 Currently available Isotope ratio PRMs from NPL

Table 3 provides information on the standard traceable isotope ratio PRMs currently available from NPL for greenhouse gases.

Table 3. Traceable isotope ratio PRMs currently available from the NPL.

Component	Amount fraction range [μmol mol^{-1}]	Typical amount fraction uncertainty (k=2) [%]	Matrix	Isotopic values available [‰]	Isotope ratio traceability
CO_2	400 – 420, 800	0.2	Synthetic air	$\delta^{13}C$: $\delta^{18}O$: +2 -8 -42 -36 -20 -22 -9 -15	VPDB
CH_4	1.8 – 2.5	0.2	Synthetic air	$\delta^{13}C$: $\delta^{2}H$: -39 -195	VPDB, VSMOW/SLAP
N_2O	0.32 – 0.35	0.5	Synthetic air	$\delta^{15}N^{SP}$: $\delta^{15}N$: $\delta^{18}O$: 1 0.2 39 -4 53 103 -0.7 16 55 20 33 40	Air N_2, VSMOW/SLAP

Further amount fractions and isotope ratios may be available on request at gas_enquiries@npl.co.uk and online at https://www.npl.co.uk/products-services/gas/climate-monitoring The webpage will be updated with new products and services as they become available. The isotope ratio values provided for N_2O and CH_4 reference materials are those of the pure gases used to prepare the PRMs. Validated isotope ratio of ambient amount fraction PRMs traceable to VPDB are available for CO_2 with typical uncertainties of 0.2 -0.5 ‰ (k=2).

5 Summary

Large volume PRMs with traceability to the SI for amount fraction and to existing scales for isotope ratio have been developed at NPL for use as calibration standards for greenhouse gas monitoring. The NPL-2023-CO_2 scale will enable the provision of amount fraction CO_2 PRMs with equations to link the amount fraction to the WMO-CO2-X2019 scale. This will improve accessibility to traceability and ensure redundancy, help to monitor long term drift in existing scales and ensure international traceability is not dependent on one institute.

Further developments aim to reduce uncertainties to achieve the network comparability goals set by the WMO-GAW for amount fraction and isotope ratio and extend the range of isotope ratios available to customers. This work lays the foundations for the establishment of an international metrology support infrastructure for enhancing data quality and comparability of the amount fraction and isotope ratio of greenhouse gas measurements and enable to identification of sources and sinks of GHG emissions.

The authors would like to acknowledge Heiko Moossen and Harro Meijer for providing isotopic measurements as part of the EMPIR project 19ENV05 STELLAR and Joachim Mohn for providing N_2O reference materials and isotopic measurements of N_2O as part of the EMPIR project 16ENV06 SIRS and via the ATMO ACCESS project. The Authors acknowledge Euramet funding via EMPIR projects 16ENV06 SIRS, 19ENV05 STELLAR, 21GRD04 isoMET, MetHIR 24GRD07 and Met CTG 24GRD06. This work was, partly funded by U.K. National Metrology System (NMS).

All authors contributed to the experimental work or discussions of the work detailed here and in the writing of this publication.

References

[1] M. J. T. Milton and T. J. Quinn, Primary methods for the measurement of amount of substance, *Metrologia,* **38** 289, (2001), DOI: 10.1088/0026-1394/38/4/1

[2] Richard J C Brown *et al.*, On the traceability of gaseous reference materials, *Metrologia,* **54**, 3, (2017), DOI: 10.1088/1681-7575/aa6ede

[3] BIPM (2021). *CCQM-GAWG Strategy document 2021-2030.*, https://www.bipm.org/documents/20126/2071059/CCQM+Strategy.pdf/31283069-94f4-f2c7-bbfc-7d652c9b3de8

[4] E. Flores *et al.*, CCQM-K120 (Carbon dioxide at background and urban level), *Metrologia*, **56**, (2019), DOI: 10.1088/0026-1394/56/1A/08001

[5] E. Flores *et al.*, International comparison CCQM-K82: methane in air at ambient level (1800 to 2200)

nmol/mol, *Metrologia* 52 08001, (2015), DOI:10.1088/0026-1394/52/1A/08001

[6] J. Viallon *et al.*, CCQM-K68.2019, nitrous oxide (N_2O) in air, ambient level, final report, *Metrologia,* **60,** 08011, (2023), DOI:10.1088/0026-1394/60/1A/08011

[7] Online, The CIPM MRA database (KCDB) What is the KCDB? https://www.bipm.org/en/cipm-mra/kcdb

[8] GAW report no 213, 17th WMO/IAEA Meeting on Carbon Dioxide, Other Greenhouse Gases and Related Tracers Measurement Techniques (GGMT-2013), https://community.wmo.int/site/knowledge-hub/programmes-and-initiatives/global-atmosphere-watch-programme-gaw/gaw-reports

[9] R. E. Hill-Pearce, *et al.,* Characterisation of gas reference materials for underpinning atmospheric measurements of stable isotopes of nitrous oxide, *Atmos. Meas. Tech.*, **14,** 5447–5458, (2021) https://doi.org/10.5194/amt-14-5447-2021

[10] K. Charoenpornpukdee, *et al.,* Step Change in Improving the Accuracy of Nitrous Oxide Reference Materials for Underpinning Atmospheric Composition Measurements, *Anal Chem,* **34**:12867-12874., (2023) DOI: 10.1021/acs.analchem.3c02160

[11] B. D. Hall *et al.,* Revision of the World Meteorological Organization Global Atmosphere Watch (WMO/GAW) CO_2 calibration scale, *Atmos. Meas. Tech.*, 14, **4**, (2021), https://doi.org/10.5194/amt-14-3015-2021

[12] Online: https://www.bipm.org/en/gas-metrology/greenhouse-gases

[13] P. J. Brewer *et al.*, SI traceability and scales for underpinning atmospheric monitoring of greenhouse gases, *Metrologia,* 55 (2018), DOI: 10.1088/1681-7575/aad830

[14] C. D. Keeling, The concentration and isotopic abundances of carbon dioxide in the atmosphere, *Tellus,* **12**, 200-203, (1960), https://doi.org/10.1111/j.2153490.1960.tb01300

[15] Online, Overview and implementation of the CIPM MRA,https://www.bipm.org/documents/20126/43742162/CIPM-MRA-P-11.pdf

[16] D. Malinovsky *et al.*, Development and characterisation of new glycine certified reference materials for SI-traceable $^{13}C/^{12}C$ isotope amount ratio measurements, *J. Anal. At. Spectrom.*, **34**, 147-159, (2019), DOI: 10.1039/C8JA00281A

[17] L. Flierl *et al.*, Absolute isotope ratios of carbon dioxide – a feasibility study, *J. Anal. At. Spectrom,* 35, 2545-2564, (2020), https://doi.org/10.1039/D0JA00318B

[18] F. Camin *et al.*, Stable Isotope Reference Materials and Scale Definitions—Outcomes of the 2024 IAEA Experts Meeting, *Rapid Commun Mass Spectrom,*39:10018.,(2025), https://doi.org/10.1002/rcm.10018

[19]Online:https://www.bipm.org/en/committees/cc/ccqm/wg/ccqm-irwg

[20] Nehrbass-Ahles, C. and Srivastava, A. and the CCQM GAWG/IRWG Joint Task Group on Stable Isotope Ratio Metrology for Atmospheric Source Apportionment (CCQM-GAWG-IRWG-TG-ISOTOP): Facilitating the development of a global measurement infrastructure for the measurement of stable isotope ratios for greenhouse gases source apportionment, EGU General Assembly (2025), EGU25-18701, https://doi.org/10.5194/egusphere-egu25-18701

[21] A. Srivastava *et al,* Developing calibration and measurement capabilities for atmospheric CH_4 stable isotope ratios at NMIs/DIs: metrology for global comparability, *Metrologia,* **62** 032001, (2025) DOI 10.1088/1681-7575/adcfa8

[22] J. Viallon *et al.,* Final report of CCQM-P204, comparison on CO_2 isotope ratios in pure CO_2, *Metrologia,* **60** 08026, (2023), DOI: 10.1088/0026-1394/60/1A/08026

[23] Online, https://www.bipm.org/en/gas-metrology/greenhouse-gases

[24] R. Hill-Pearce *et al.*, Synthetic Zero Air Reference Material for High Accuracy Greenhouse Gas Measurements, *Anal. Chem.*, 91, **3**, 1974-1979 (2019) DOI: 10.1021/acs.analchem.8b04297

[25] Online, https://www.npl.co.uk/products-services/environmental/the-boreas-preconcentrator

[26] C. Rennick *et al.,* Boreas: A Sample Preparation-Coupled Laser Spectrometer System for Simultaneous High-Precision In Situ Analysis of $\delta^{13}C$ and $\delta^{2}H$ from Ambient Air Methane, *Anal. Chem.*, 93, **29**, 10141–10151, (2021), https://doi.org/10.1021/acs.analchem.1c01103

[27] H. Nara *et al.*, Effect of air composition (N_2, O_2, Ar, and H_2O) on CO_2 and CH_4 measurement by wavelength-scanned cavity ring-down spectroscopy: calibration and measurement strategy, *Atmos. Meas. Tech.,* **5**, 2689–2701, (2012), DOI:10.5194/amt-5-2689-2012

[28] E. Flores, *et al.,* Towards Standardized Whole Air and Purified Air Gas Production and Characterization for Greenhouse Gas Standards, In proceedings of metrology for climate action 2022, https://www.bipmwmo22.org/submissions/submission/13#:~:text=Gas%20standards%20required%20by%20the,produced%20using%20an%20air%20matrix

[29] Online, https://www.bipm.org/en/-/2021-04-01-co2-tg

[30] P. J. Brewer, *et al.,* Breakthrough in Negating the Impact of Adsorption in Gas Reference Materials, *Anal. Chem.*, 91, **8,** (2019), https://doi.org/10.1021/acs.analchem.9b00175

[31] R. Hill-Pearce *et al.* Primary gas reference materials for underpinning measurements of $\delta^{13}C$-CO_2 and $\delta^{18}O$-CO_2 with traceability to the VPDB scale [submitted]

[32] J. Mohn, *et al.*, Isotopically characterised N_2O reference materials for use as community standards, *Rapid Commun Mass Spectrom*. 36, **13**, e9296, (2022). DOI:10.1002/rcm.9296

[33] C. Rennick *et al.*, Good practice guide for accurate methane isotope ratio measurements using laser spectroscopyhttps://empir.npl.co.uk/stellarproject/wp-content/uploads/sites/62/2023/12/19ENV05-STELLAR-D6-v1.pdf

[34] E. Safi *et al.*, Interlaboratory Comparison of $\delta^{13}C$ and δD Methane Measurements in Synthetic Air for Data Harmonization, (not yet published).

Characterization of HCl Primary Reference Gas Standards for Emission Monitoring

Iris de Krom [1*], *Sophie* Schout[1], *Luuk* Meijer[1], *Evtim* Efremov[1], *Stefan* Persijn[1], *Gerard* Nieuwenkamp[1] and *Adriaan* van der Veen[1]

[1] Van Swinden Laboratorium B.V. (VSL), Thijsseweg 11, 2629 JA Delft, Netherlands

Abstract. Hydrogen chloride (HCl) is a key acidic pollutant emitted by waste-to-energy plants, as well as in industrial processes such as cement manufacturing and fossil fuel combustion. Due to its harmful impact on human health and the environment, strict emission limits are enforced under regulations like the European Industrial Emissions Directive (2010/75/EU), revised in 2024, which require regular monitoring of HCl. In Europe, HCl emissions are typically monitored using wet-chemical methods in accordance with EN 1911. These methods exhibit high uncertainty at concentrations below 10 mg m^{-3}, the EU's daily average emission limit, prompting the exploration of more accurate alternatives. Optical gas analysers, already approved by the U.S. Environmental Protection Agency, offer a promising alternative with lower measurement uncertainty. These instruments require calibration with reference gas standards of established metrological traceability. To meet this requirement, VSL employed an optical gas analyser to investigate key factors influencing the development of primary reference materials (PRMs) with HCl. The study focused on stability, adsorption behaviour, and pressure dependence of HCl PRMs. As a result, VSL successfully produced PRMs for HCl in high-pressure cylinders, achieving amount fractions as low as 5 µmol mol^{-1} in nitrogen (approximately 10 mg m^{-3}) with an expanded relative uncertainty of 3% (k = 2). These PRMs are essential for ensuring accurate and reliable calibration of optical gas analysers used in emission monitoring applications.

1 Introduction

Hydrogen chloride (HCl) is a significant contributor to acid pollution, originating from waste-to-energy facilities and industrial activities such as cement production and fossil fuel combustion. Its detrimental impact on human health and ecosystems has led to strict regulatory controls, notably under the European Industrial Emissions Directive (2010/75/EU), which was revised in 2024 to reinforce emission limits and monitoring requirements [1]. Compliance with these regulations typically relies on wet-chemical measurement techniques, as prescribed by EN 1911 [2]. These methods exhibit considerable uncertainty at low concentrations however, particularly near the daily average limit of 10 mg m^{-3}, creating challenges for accurate monitoring.

To address these limitations, optical gas analysers have emerged as a promising alternative, offering improved precision and already gaining approval from the U.S. Environmental Protection Agency. Nevertheless, their effective deployment depends on calibration using reference gas standards with robust metrological traceability. Developing such standards for HCl presents unique challenges due to its reactive nature and adsorption behaviour, especially at low amount fraction levels [3-5]. To evaluate and benchmark the competence for HCl gas standards and analysis the first key comparison (CCQM-K175) was executed between 2020 and 2024 [6]. For the CCQM-K175, travelling gas standards with approximately 30 µmol mol^{-1} HCl in nitrogen were used. During the key comparison, VSL utilised HCl primary gas standards with mole fractions from 100 µmol mol^{-1} to 5 µmol mol^{-1} for the calibration of their analyser [6]. Building on the successful results obtained in the CCQM-K175 key comparison, VSL has undertaken further investigation into the properties of our HCl gas standards. This study focussed on the critical factors influencing the preparation of primary reference materials (PRMs) for HCl, including stability assessment, cylinder wall adsorption, and pressure dependence of the amount fraction HCl. The results obtained in the CCQM-K175 and this study demonstrate the successful production of PRMs in high-pressure cylinders at amount fractions as low as 5 µmol mol^{-1} in nitrogen [6]. These advancements provide a solid foundation for reliable calibration of optical gas analysers, supporting accurate emission monitoring and regulatory compliance.

2 Methods and materials

2.1 Materials for preparing HCl PRMs

HCl gas is highly corrosive and toxic, making the accurate and safe preparation of HCl gas mixtures critically important. To ensure accurate and safe preparation, a dedicated gas filling station was used to prepare static gas mixtures in high-pressure cylinders starting from pure HCl gas. The facility, custom-built in collaboration with Nippon Gases Belgium N.V. and HTR Rotterdam B.V. (Figure 1), was commissioned in 2024. This facility not only ensures a safer working environment but also enables the production of highly accurate static gas mixtures in accordance with ISO 6142-1, with a preparation expanded uncertainty of just

* Corresponding author: idekrom@vsl.nl

0.08% ($k = 2$) [7]. The gas filling station is specifically designed to sample acid-forming gases, such as HCl, from a gas cylinder. It reduces the cylinder pressure to an outlet pressure with an needle valve. The design allows purging to be carried out with an inert gas (e.g., nitrogen), so it can be used with reactive, toxic and corrosive gases and gas mixtures of middle to high amount fractions. Purging with an inert gas, before initial start-up or while the cylinders are being changed, flushes all the atmospheric air from the system to ensure safe utilization of the process gas. The gas filling station is primarily constructed of electropolished stainless steel.

Fig. 1. Dedicated HCl gas filling station.

For mixtures with HCl amount fractions greater than or equal to 1000 µmol mol^{-1} 10 L stainless steel high-pressure cylinders from Nippon Gases (BE) are used. Amount fractions below 1000 µmol mol^{-1} HCl are prepared in 10 L aluminium high-pressure cylinders from Luxfer (UK) with a special HCl passivation from Air Liquide (FR).

Pure HCl gas (quality 5.0, >99.999%) was purchased from Nippon Gases (BE). The purity table provided by the supplier was adopted (Table 1). The pure nitrogen (grade 6.0, >99.9999%, BIP+, Air Products) used for the dilutions was checked for impurities in accordance with ISO 19229 [8, 9].

Table 1. Purity table HCl.

Component	Amount fraction (µmol mol^{-1})	Uncertainty (µmol mol^{-1})
Methane	0.50	0.29
Carbon monoxide	0.50	0.29
Carbon dioxide	1.0	0.6
Water	0.50	0.29
Nitrogen	1.5	0.9
Oxygen	0.50	0.29
Hydrogen chloride	999995.5	1.2

2.2 Gravimetric preparation of HCl PRMs

Two series of HCl PRMs were prepared, using the new gas filling station, starting from pure HCl by dilution with nitrogen (Figure 2). The preparation process complies with ISO 6142-1, which specifies the gravimetric method for producing calibration gas mixtures with high accuracy and traceability [7, 10].

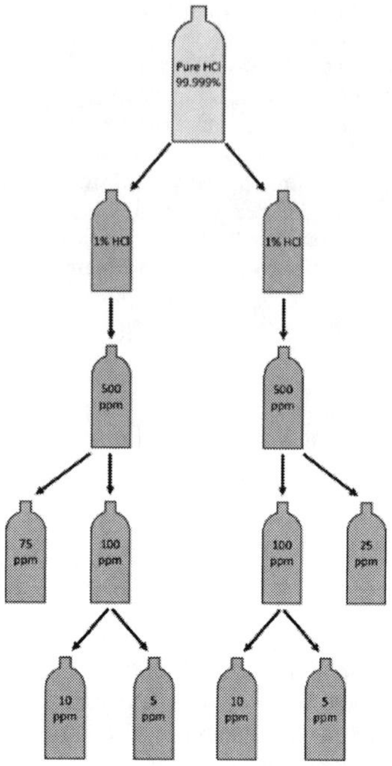

Fig. 2. Series of HCl PRMs prepared.

2.3 Analyser for HCl gas mixtures

For the analysis of the HCl PRMs, a spectrometer based on optical feedback cavity enhanced absorption spectroscopy (OF-CEAS) from AP2E (FR) was used. The cylinders were equipped with a Silconert coated stainless steel pressure regulator and the regulator was flushed prior to use. The measurements are conducted manually by connecting the PRMs to the analyser using short pieces of PFA tubing. A flow of 1000 ml min^{-1}, controlled by a Silconert coated Bronkhorst (NL) thermal mass flow controller, is led to the analyser.

The analyser can be used in the range from 1 µmol mol^{-1} up to 350 µmol mol^{-1}, with a detection limit of 0.6 µmol mol^{-1} and an expanded measurement uncertainty of 1.6% ($k = 2$), relative. Under dry conditions, the response time of the analyser is between 15 minutes and 60 minutes. The analyser is calibrated with at least 3 primary gas standards for the verification of new PRMs. As a calibration function, a straight-line model was used, established according to ISO 6143 [11].

2.4 Stability study

An important property of reference materials is the stability period, reflected as an expiration date on the certificate [12]. In the case of the static PRMs, this is the period during which the composition of the gas mixture in a cylinder is stable within the range of the measurement uncertainty. A stability study on 10 µmol mol^{-1} and 1 µmol mol^{-1} HCl PRMs was performed with 2 different cylinder passivation treatments. One general passivation and one special HCl passivation (Figure 3). To two of the 1 µmol mol^{-1} PRMs, 2 µmol mol^{-1} of water was added.

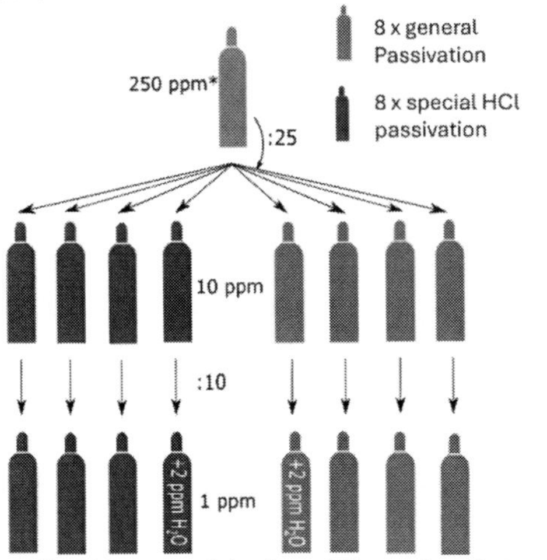

Fig. 3. HCl PRMs prepared for the stability study with two different cylinder passivation treatments.

The 10 µmol mol^{-1} PRMs were analysed five times over a period of more than 500 days, the 1 µmol mol^{-1} PRMs were analysed three times over a period of almost 400 days. Before each analysis the analyser was calibrated with primary gas standards.

2.5 Adsorption behaviour

HCl is very reactive and adsorbs easily on surfaces including the cylinder wall, reducers and tubing. Adsorption can lead to losses during and after preparation. To determine whether this loss is significant the decanting method is typically used [13, 14]. For this test, two 10 µmol mol^{-1} PRMs were used. Approximately half of the gas was decanted into an empty cylinder followed by cross-comparison.

2.6 Pressure dependence

Finally, the influence of pressure on the composition of HCl PRMs was determined. It is important that the amount fraction HCl remains within the coverage interval stated on the certificate when the pressure in the cylinder decreases due to use of the PRM. HCl strongly adsorbs onto the internal surface of the cylinder. As the mixture is consumed, the pressure in the cylinder reduces and the amount faction of the HCl can increase substantially [4].

To determine this, a 10 µmol mol^{-1} PRM was analysed starting at a pressure of 7 MPa in the cylinder. At 3 moments in time the measurement was stopped to determine the pressure in the cylinder. After 2 days of analysis, the cylinder was empty.

3 Results and discussion

3.1 Development HCl PRMs

After preparation of the two sets of HCl PRMs, the composition of the gas standards was verified. Verification took place against existing primary gas standards that were also employed for the CCQM-K175 measurements [7]. The new series of PRMs showed good agreement with each other and were aligned with the primary gas standards used for the CCQM-K175 within the expanded uncertainty of 3% ($k = 2$) used for CCQM-K175 (Figure 4 and Figure 5).

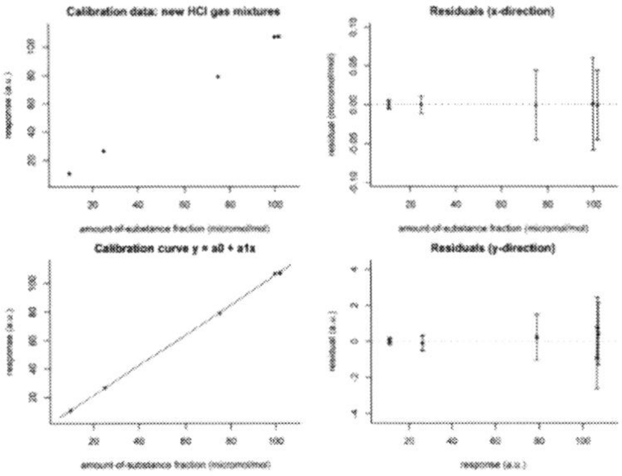

Fig. 4. Linear regression function and residuals for the new HCl PRMs.

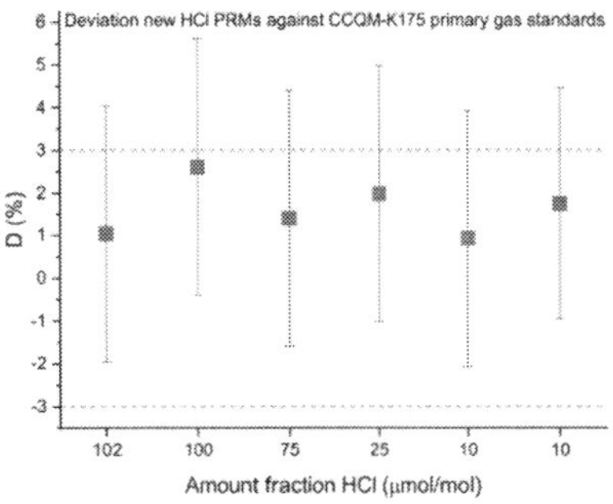

Fig. 5. Deviation of the new HCl PRMs compared to the calibration curve with the CCQM-K175 primary gas standards.

For a satisfactory fit of the data, it is required that the absolute values of the normalized residuals do not exceed 2. The normalized residual is the residual divided by the standard uncertainty [11]. All absolute values of the normalized residuals for new HCl PRMs were smaller than 2. The relative deviations of the new HCl PRMs compared to the calibration curve with the CCQM-K175 primary gas standards are within ± 3.0%.

3.2 Stability study

For the stability study 4 × 4 HCl PRMs were prepared. In Figure 6 the results of the stability study at 10 µmol mol^{-1} HCl are shown and in Figure 7 of the 1 µmol mol^{-1} HCl PRMs.

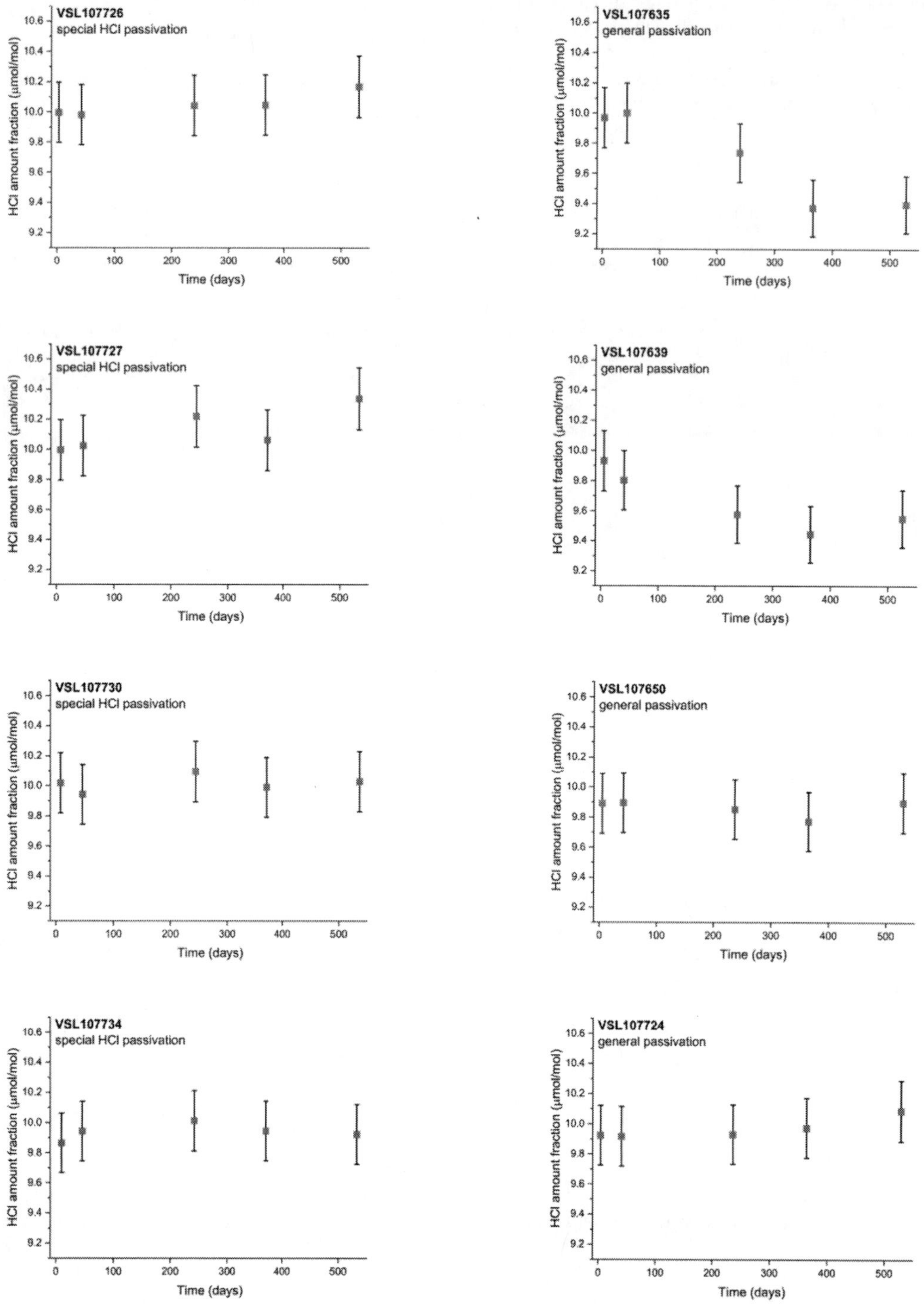

Fig. 6. Results stability study of 10 µmol mol^{-1} HCl. Fraction plotted against the time, uncertainty bars indicating 2.0% stability.

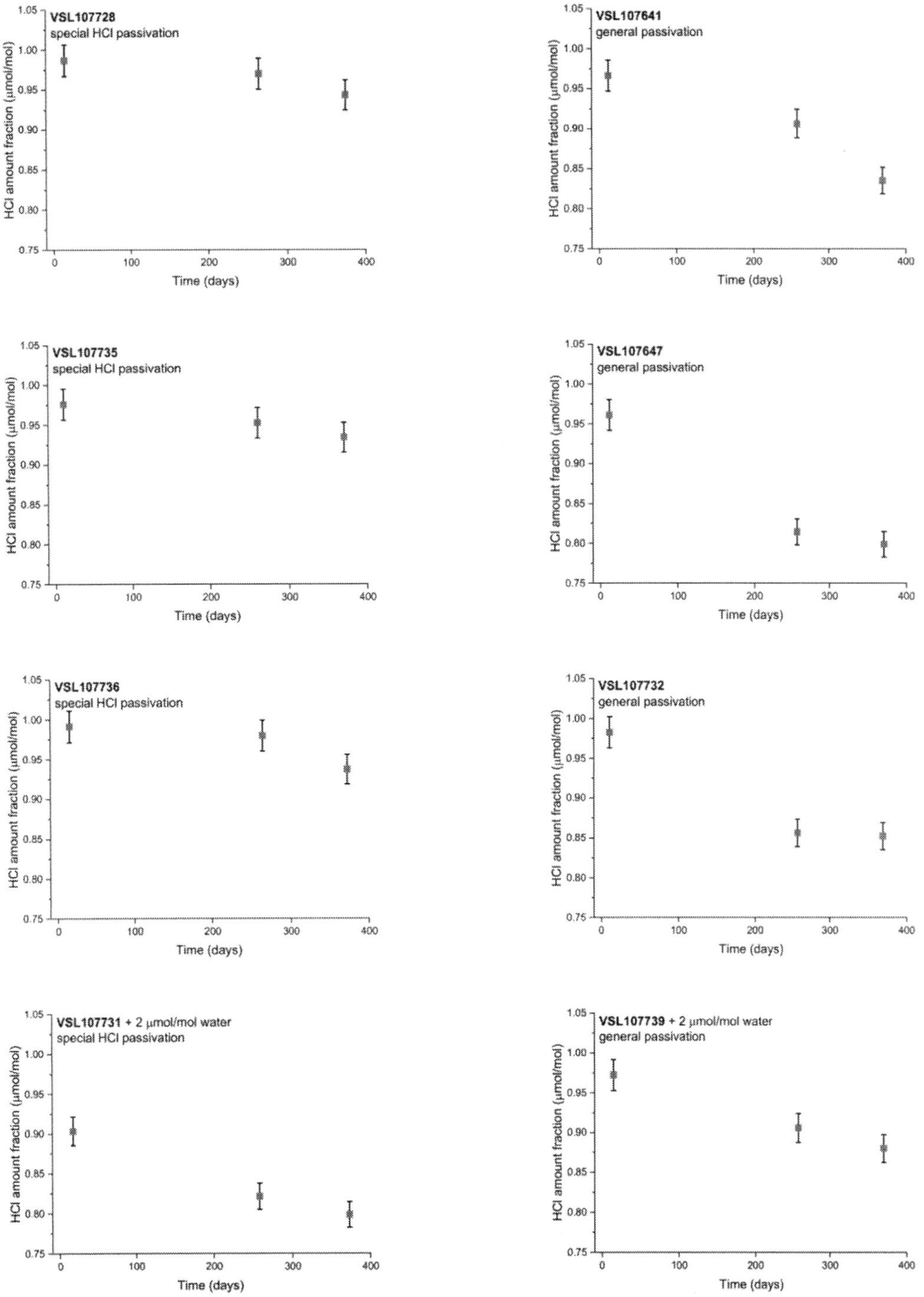

Fig. 7. Results stability study of 1 μmol mol^{-1} HCl. Fraction plotted against the time, uncertainty bars indicating 2.0% stability.

The results show excellent stability over a period of 500 days for the 10 μmol mol^{-1} PRMs in cylinders with a special HCl passivation. Based on the stability data for these mixtures, a 1-year stability within ± 2.0% was determined using regression to assess the data for meaningful instability and a random effects model to evaluate the uncertainty [12]. The calculations were performed in R software.

Two of the 10 µmol mol⁻¹ PRMs in cylinders with a general passivation were found to be unstable (VSL107635 and VSL107639).

The 1 µmol mol⁻¹ PRMs in cylinders with a special HCl passivation are stable for 200 days. Addition of water (VSL107731) leads to higher losses of HCl and instability of the mixture. The PRMs prepared in cylinders with a general passivation are not stable and addition of water (VSL107739) does not have an influence on the stability, this could indicate that these cylinders already contain water before filling, leading to instability.

3.3 Adsorption behaviour

Two 10 µmol mol⁻¹ mixtures were prepared (VSL121486 and VSL207730) approximately half of both gas mixtures was decanted into empty cylinders (VSL121490 and VSL122544). All four mixtures were analysed three times (Figure 8).

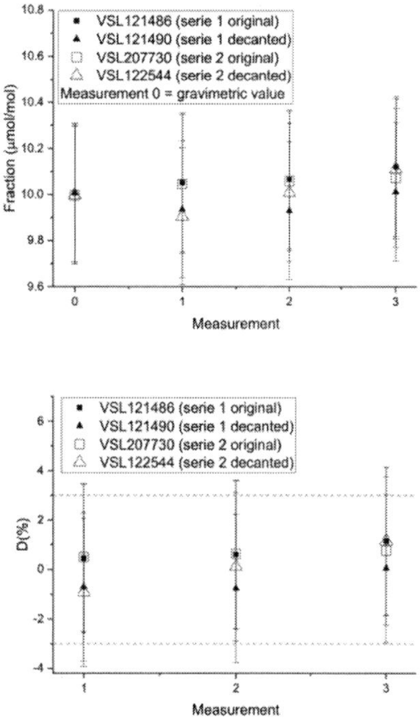

Fig. 8. Top – Fraction of the 10 µmol mol⁻¹ mixtures compared to a calibration curve with HCl PRMs. Bottom – Relative deviation of the 10 µmol mol⁻¹ mixtures compared to the calibration curve.

The decanted mixtures (VSL121490 and VSL122544) have a lower HCl fraction and more negative deviation compared to the original mixtures, except during measurement 3 where the fraction of VSL122544 is slightly higher compared to the original mixture (VSL207730). The maximum deviation is -1.36% (VSL121490) and -1.41% (VSL122544). This deviation falls well within the 3.0% uncertainty, therefore the loss is insignificant.

3.4 Pressure dependence

VSL307728 was analysed to determine the pressure dependence of the amount fraction HCl when emptying the gas cylinder (Figure 9).

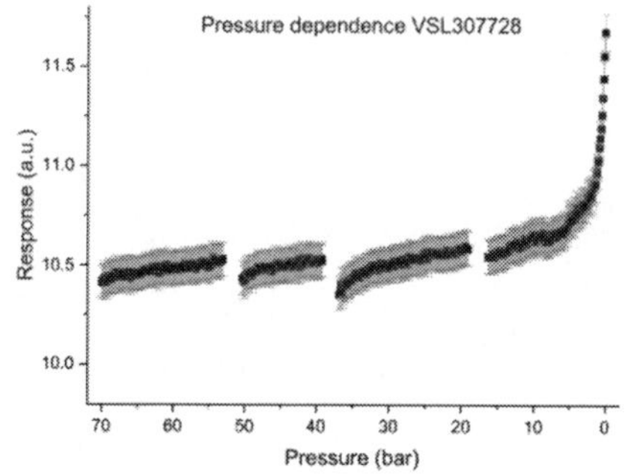

Fig. 9. Pressure dependence of the amount fraction HCl of a 10 µmol mol⁻¹ gas standard. Gaps indicate pressure measurement. The dark points indicate the measurement values, the light colour indicates the expanded uncertainty of 3%.

In the beginning, and after pressure measurements, the analyser signal had to stabilize. After stabilisation the signal is stable until a pressure of 0.5 MPa is reached. Based on the results the gas mixtures cannot be used below a pressure of 0.5 MPa. This is below the default limit of 1 MPa for all PRMs and certified reference materials of VSL. The increase in fraction below 0.5 MPa is up to 10%. This effect can be attributed to the strong interaction of HCl with the cylinder surface compared to other gases [4].

3.5 HCl PRM uncertainty

For CCQM-K175 a relative expanded uncertainty of 3% (k = 2) was calculated. Based on the results obtained in this study the uncertainty budget for the HCl PRMs was determined with sources from gravimetry, measurement uncertainty, stability and adsorption (Table 2).

Table 2. Uncertainty budget HCl PRMs

	U_{rel} (%)
Gravimetric uncertainty	0.08
Measurement uncertainty	1.6
Stability	2.0
Adsorption	1.4
HCl PRM	**3**

The relative expanded uncertainty for the HCl PRMs down to 5 µmol mol⁻¹ is 3% (k = 2).

4 Conclusion

PRMs for HCl in high-pressure cylinders were successfully developed and characterised. Stability, adsorption and pressure dependence studies confirm the robustness of these mixtures, supporting the claimed expanded uncertainty of 3%. This work is underpinned by the excellent comparison with the primary gas standards used for the CCQM-K175, allowing the continuation of the metrological traceability chain. These PRMs are critical for the traceable calibration of optical gas analysers in emission monitoring, ensuring accurate, reliable, and comparable measurements across laboratories and industries.

Acknowledgments

The research described in this paper received funding from the Ministry of Economic Affairs of the Netherlands. Some of the research was carried out within the EMPIR (European Metrology Programme for Innovation and Research) project 17IND09 MetAMCII (under Euramet), co-financed by the Participating States and from the European Union's Horizon 2020 research and innovation programme.

ORCID iDs

Iris de Krom https://orcid.org/0000-0001-5895-5959
Luuk Meijer https://orcid.org/0009-0001-5783-3375
Stefan Persijn https://orcid.org/0000-0002-4356-4266
Adriaan van der Veen https://orcid.org/0000-0002-9648-5123

References

1. Directive 2010/75/EU of the European Parliament and of the Council (2010). https://eur-lex.europa.eu/eli/dir/2010/75/2024-08-04
2. EN 1911 Stationary source emissions – Determination of mass concentration of gaseous chlorides expressed as HCl – Standard reference method (2010).
3. J. Jung, B. Kim, S Oh, Development of 100 μmol mol^{-1} HCl primary standard gas mixtures in high-pressure cylinders, Metrologia, **56**, 015007 (2019). http://doi.org/10.1088/1681-7575/aaf2b1
4. P.J. Brewer, R.J.C. Brown, K.V. Resner, R.E. Hill-Pearce, D.R. Worton, N.D.C. Allen, K.C. Blakley, D. Benucci, M.R. Ellison, Influence of pressure on the composition of gaseous reference materials, Anal. Chem., **90**, 3490 (2018). http://doi.org/10.1021/acs.analchem.7b05309
5. J.I.T. van Wijk, G. Nieuwenkamp, T.L. Hafkenscheid, Reference materials at emission levels of NH_3 and HCl, Fresenius J Anal Chem, **370**, 200 (2001). https://doi.org/10.1007/s002160100825
6. J. Jung, S. Oh, B. Kim, C. Cecelski, C. Goodman, J. Carney, D.R. Worton, Y. Hristova, M. Ward, R. Hill-Pearce. P. Brewer, J. Nwaboh, V. Ebert, L.A. Konopelko, Y.A. Kustikov, A.V. Kolobova, I.K. Chubchenko, I. de Krom, E. Efremov, A. van der Veen, International comparison CCQM-K175 – Hydrogen chloride in nitrogen, Metrologia, **62**, 08028 (2005). http://doi.org/10.1088/0026-1394/62/1A/08028
7. ISO 6142-1 Gas analysis – Preparation of calibration gas mixtures – Part 1: Gravimetric method for class I mixtures (2015).
8. ISO 19229 Gas analysis – Purity analysis and the treatment of purity data (2019).
9. A.M.H. van der Veen, E. T. Zalewska, J.I.T. van Wijk, M. Kobayashi, D. Akima, S. Uehara A.L. Fioravante, C.R. Augusto, CC. Ribeiro, V. Silva, F. Dias, A. Botas, C. Costa, J. Lee, J. Lee, J. Lim, H.K. Bae, N. Kang, C.E. Cecelski, K.J. Harris, W.R. Miller, J. Carney, J. Tshilongo, N.G. Ntsasa, M.I. Jozela, N. Leshabane, P. Mohweledi Marebane, D.R. Worton, E.B. Mussell Webber, S. Moreno, P.J. Brewer, L.A.K Konopelko, A.V. Kolobova, V.V. Pankratov, O.V. Efremova, International comparison CCQM-K3.2019 automotive exhaust gases, Metrologia, **60**, 08008 (2023). http://doi.org/10.1088/0026-1394/60/1A/08008
10. M.J.T. Milton, G.M. Vargha, A.S. Brown, Gravimetric methods for the preparation of standard gas mixtures, Metrologia, **48,** R1 (2011). http://doi.org/10.1088/0026-1394/48/5/R01
11. ISO 6143 Gas Analysis – Comparison methods for determining and checking the composition of calibration gas mixtures (2025).
12. A. van der Veen, T. Linsinger, A. Lamberty, Uncertainty calculations in the certification of reference materials. 3. Stability study, Accred Qual Assur **6**, 257-263 (2001). https://doi.org/10.1007/s007690000292
13. S. Lee, M.E. Kim, S.H. Oh, J. S. Kim, Determination of physical adsorption loss of primary standard gas mixtures in cylinders using cylinder-to-cylinder division, Metrologia, **54,** L26 (2017). http://doi.org/10.1088/1681-7575/aa87b8
14. S.T. Persijn, A. Baldan, A new look at the sorption kinetics in reference gas standards, Meas. Sci. Technol., **34,** 115018 (2023). http://doi.org/10.1088/1361-6501/ace9ee

AUTHOR INDEX